Springer Tracts in Electrical and Electronics Engineering

Springer Tracts in Electrical and Electronics Engineering (STEEE) publishes the latest developments in Electrical and Electronics Engineering - quickly, informally and with high quality. The intent is to cover all the main branches of electrical and electronics engineering, both theoretical and applied, including:

- Signal, Speech and Image Processing
- Speech and Audio Processing
- Image Processing
- Human-Machine Interfaces
- Digital and Analog Signal Processing
- Microwaves, RF Engineering and Optical Communications
- Electronics and Microelectronics, Instrumentation
- Electronic Circuits and Systems
- Embedded Systems
- Electronics Design and Verification
- Cyber-Physical Systems
- Electrical Power Engineering
- Power Electronics
- Photovoltaics
- Energy Grids and Networks
- Electrical Machines
- Control, Robotics, Automation
- Robotic Engineering
- Mechatronics
- Control and Systems Theory
- Automation
- Communications Engineering, Networks
- Wireless and Mobile Communication
- Internet of Things
- Computer Networks

Within the scope of the series are monographs, professional books or graduate textbooks, edited volumes as well as outstanding PhD theses and books purposely devoted to support education in electrical and electronics engineering at graduate and post-graduate levels.

Review Process
The proposal for each volume is reviewed by the main editor and/or the advisory board. The books of this series are reviewed in a single blind peer review process.

Ethics Statement for this series can be found in the Springer standard guidelines here https://www.springer.com/us/authors-editors/journal-author/journal-author-helpdesk/before-you-start/before-you-start/1330#c14214

Pranab Kumar Dey

Power Supplies for Synchrotron Accelerators

Springer

Pranab Kumar Dey
Rajarhat, West Bengal, India

ISSN 2731-4200 ISSN 2731-4219 (electronic)
Springer Tracts in Electrical and Electronics Engineering
ISBN 978-981-97-7605-4 ISBN 978-981-97-7606-1 (eBook)
https://doi.org/10.1007/978-981-97-7606-1

This Springer imprint is published by the registered company Springer Nature Singapore Pte Ltd.
The registered company address is: 152 Beach Road, #21-01/04 Gateway East, Singapore 189721, Singapore

Dedicated to my parents, Late Renuka Dey and Late Manmatha Nath Dey, for their affection and blessings; to my wife, Sukla, for her love, inspiration, and continuous support during the arduous task of writing this manuscript; and to my daughters, Dolonchampa and Dwaita, for constantly encouraging me to unfold the glorious experiences of the past.
Pranab Kumar Dey

Preface

More than 100 years back from now, results of Lord Rutherford's epoch-making experiments, his achievements and scientific contributions in atomic scattering in 1906, the theory of atomic structure with the disintegration of nuclei in 1911, pioneering findings on radioactivity in 1919, and many such discoveries acknowledged by the scientific community over a long period encompassing large workstations including Cavendish Laboratory, Cambridge, England, made history. It was those glorious years when the interest in accelerating particles with sufficient energy had its beginning. Ernest Rutherford, being the first one to split the atom, was the central figure in the study of radioactivity, which led to the present advancement of nuclear physics in different directions. In the course of time, it was established that acceleration of the charged particles inside a vacuum tube depends on the strength of the electric field developed by the electric potential applied across the terminals of the electrodes. The advanced research on nuclear physics is attributed to the relentless efforts of the scientists and technologists over the decades, which went on opening up new avenues for investigations, made continuous attempt to upgrade technologies to match the requirements of new concepts, meet the challenging tasks of designing particle accelerators to raise energy levels, and improve performance to push forward the goal towards finding the ultimate particle. In the early 1930s, it was possible to accelerate ions to energies of few hundred KeV only, by applying direct voltage because the operation of these accelerators was restricted to low energy level due to voltage breakdowns. The consistent demand of higher energies led the researchers to concentrate in finding suitable schemes to continue their experiments with particles accelerating at higher energies. But how minimum energies of 500 KeV or more could be developed to accelerate the charged particle to disintegrate nuclei was of serious concern to the investigators. The testimony to such spectacular advancement made over the years from atomic physics to nuclear physics and how initial energy level has been raised to the present stage of operation is readily available from various literatures including well-known natural progression chart prepared by Livingston, which illustrates the evolution of various types of accelerator with time. It was a global attempt by the scientists jointly with the theoretical physicists, experimental physicists, and technologists to put enormous effort to reach the set goal. The grandeur of this journey is full of excitement and thrills to the technologists in every stages of design, development, fabrication, manufacturing, installation, commissioning, testing, and operation. The progress of science, engineering, and technology goes hand in hand and plays important roles in contributing their share for carrying forward the costly research activities in different fields with the expectation that the fruits of such research findings will be paid back to the welfare of the society and ultimate benefit to the mankind. Working in accelerator complex is a great challenge to the scientists and engineers where frequent temptations to cross the boundaries of unknown territories in sorting out different types of technical limitations are a regular feature unlike some routine assignments. Technical literatures on particle accelerators are readily available, but working at "Rutherford High Energy Laboratory" and sharing the valued experience gained from the most reputed nuclear physics center in England, where a class of subnuclear particles were created in energetic collisions between protons from the high-energy machine and target nuclei, were somewhat different. The disintegrated particles produced as a result of such bombardments were analyzed to find the detailed mechanisms of holding the nucleus together. It will give immense pleasure to me to share the unique experience that I gained while working in such an environment with others, interested to know the technical activities inside the boundaries of such organizations. This is a tribute particularly to the Science Research Council, UK, for accepting me in their fold when the research activities at RHEL were picking up to meet the demands of the scientists. To a large extent, I also owe to the experts of other nuclear physics laboratories and acknowledge my sincere gratitude to them who continue to enrich me with their latest research findings on the subject. The world of accelerator physics, magnet physics, and particle accelerators is vast, and numerous operating accelerators with different technologies adopted to impart required energies for their facilities need some discussion. This book only tried to hold the core of the subject by describing few significant practical achievements in the application of electrical engineering equipment, power electronics, and engineering systems, specifically designed, selected, and developed for particle accelerators. Importance for improving the performance and operation of accelerator is a crucial factor for which course corrections required during various stages of design and operation are highlighted within the limited space. It has cut short the description of the vast subject and concentrated on power supplies by retaining only the cardinal points. The book briefly deals with the operational difficulties faced due to technical limitations that surfaced during different stages of development to show how gracefully the problems were solved to push forward the development of particle accelerators from low energy at the initial stages to operation at higher energy levels. Attempt has been made to emphasize the need of revisiting the subject and analyze the different possibilities of power supplies incorporated as a routine evaluation procedure in the system including the latest state of the art before finalizing the solution. Interested engineers can assess how the process of evaluations was made before selecting a suitable technology and the stresses involved in finding a long-term solution for performance improvement. Practicing engineers may be interested to know the operational records of the earlier accelerators and take notes on how the critical decisions were taken at the preliminary stages of design to avoid the future risk of accepting alteration in the technical demand afterwards, which may incur a serious impact on the financial loading for a costly project of this scale and affect the time-bound program. The complexities of the subject have been toned down by avoiding intricacies of theories for ease of presentation as far as possible considering the choice of large number of readers. It covers only the basics of power supplies for particle accelerator operation. The perception in analysis, mode of presentation, and explanation may vary, for which suggestion will always be appreciated.

Rajarhat, West Bengal, India Pranab Kumar Dey

Contents

1 **Initial Accelerators** 1

2 **Power Supplies for High Energy Particle Accelerators of 1950s** 9

3 **Comparison of Power Supply Equipment of Synchrotrons of 1950s** 35

4 **Changing Patterns of Magnet Excitations and Energy Storage Schemes for Particle Accelerators** 47

5 **Semiconductor Devices, Converter Topologies, and Control System Strategies for Accelerators** 85

6 **Miniature Model of Developing Functional Structure of Accelerator Project** 105

7 **Review of Power Supplies to Synchrotron Accelerators After 1960s** 117

About the Author

Pranab Kumar Dey graduated with a degree in electrical engineering from Faraday House Engineering College, London, and joined Rutherford High Energy Laboratory, Berkshire, England, as an engineer. He left Rutherford High Energy Laboratory to pursue postgraduate studies at the University of Nottingham after receiving a scholarship from the Berkshire County Council. Unfortunately, he returned to India without completing his course. He then joined the Saha Institute of Nuclear Physics in Calcutta. His keen interest in power electronics and control engineering led him to join Hindustan Copper Ltd, an integrated copper complex, where he could find different applications of power electronics in mining and process controls in plants.

Chapter 1
Initial Accelerators

It all started hundred years back from now when the path-breaking results of the fundamental research of Lord Rutherford, J.J. Thompson, and others sparked the future possibilities of atomic physics from 1919. Today, we acknowledge our deepest gratitude to the torchbearers and front-runners, who guided us throughout to draw the road map of developing particle accelerators in stages to reach the spectacular height over the years which will take endless pages to record and describe their dedication and appreciate their achievements. The description of primary part about particle accelerator is available in textbooks to strengthen the foundation and for conscientious readers the reports on the sequence of events leading to generating higher voltages and subsequent progress of developing higher energies for initial particle accelerators are available in various technical literatures. However, to refresh the stages of such developments which continued at a very slow pace from 1920 to end of World War II are highlighted in brief to emphasize the enormity of the task and the hurdle faced from the beginning by the most talented scientists and engineers to all who are interested to know little more and identify the path that led to the right direction to reach at the present stage. At the beginning, it will be appropriate to acknowledge once again, our gratitude to hundreds of prominent scientists and engineers for their contribution, dedication, relentless effort, and guidance in different fields and giving us this opportunity to share their experiences in the triumphant journeys supporting the course of advancement of particle acceleration. The evolution of initial accelerators can be compared to the invention of fire in the prehistoric era. The woods were there, but the technology of lighting the woods was not known. Three path-breaking equations expressed below bounded and distinctly defined in general forms considered fundamental to the design of particle accelerators were well established (Bryant n.d.) but transforming the theoretical equations into reality and developing the practical concept of acceleration came after a long struggle.

P. K. Dey, *Power Supplies for Synchrotron Accelerators*, Springer Tracts in Electrical and Electronics Engineering,
https://doi.org/10.1007/978-981-97-7606-1_1

$$\vec{\boldsymbol{E}} = -\vec{\nabla}\varphi - \frac{\delta \vec{A}}{\delta t} \tag{1.1}$$

$$\vec{\boldsymbol{B}} = \vec{\nabla} \times \vec{\boldsymbol{A}} \tag{1.2}$$

Two separate types of particle accelerators, either static or time varying could emerge from the eq. (1.1) which reveals that acceleration of particles depends on the application of electric field. The first term is the static field derived from a scalar potential, whereas the second term a vector potential described the time-varying field which is the prime factor in the functioning of all present-day high-energy accelerators. Apart from that, when eq. (1.2) is substituted in eq. (1.1), it yields Faraday's law

$$\vec{\nabla} \times \vec{\boldsymbol{E}} = -\frac{\delta \vec{\boldsymbol{B}}}{\delta t} \tag{1.3}$$

The eq. (1.3) relates the electric field to the rate of change of magnetic field. In summing up, it is established that, with the bundle of theoretical stipulations expressed above, generation of higher energies for particle acceleration is feasible with proven technology support in generating higher voltages. In the early part of twentieth century, the researchers had to face tremendous hurdles to continue with their basic experiments on nuclear forces and elementary particles which required higher energies. At the initial stage, accelerating particles to the required energy levels appeared as serious challenge due to lack of knowledge in generating higher voltages which was the cause for concern to the experimenters. The objective was to satisfy the simple link, higher energies could be imparted to the particles only with the supply of higher voltages to the electrode terminals at the primary stages and subsequent change of modes at a later stage. This drawback to find suitable scheme of developing higher voltages inspired them to involve in vigorous search in all directions to come out from this blockage. During such initial period, a number of techniques were attempted in right directions to find few viable solutions of generating higher voltages. Prominent among them were the schemes developed by Van de Graaff, Cockcroft-Walton, Wideroe, and Lawrence whose findings and working models attributed significantly in the progress on few experiments. On the development and application of initial accelerator, the emphasis was given first on direct acceleration to utilize the principle that charged particles can be accelerated in the presence of electrostatic field when full voltage generated from high electrical potential source which do not change with time is applied between the electrodes. Thus, the only available solution for continuing research in nuclear physics at the beginning was to apply high electric potential to the electrodes of an evacuated tube for accelerating subatomic particles to higher speeds. Van de Graaff and Cockcroft-Walton, the three most eminent scientists worked on finding ways of raising higher voltages through separate routes adopting different techniques which subsequently strengthened the development of electrostatic accelerators. One such technique focused on mechanical arrangements and the others attempted in designing electrical circuits to develop voltage multiplier by raising the voltages to higher level required for particle acceleration. The apparatus design for both the techniques assumed different forms depending on the type of charged particle, electrons or protons and to the extent of energies the particles were required for acceleration.

The Van de Graaff machine is a mechanical device in which the electrical charge is transferred to the high voltage terminal on a rapidly moving belt driven by a pulley to accumulate electric charge on a hollow metal globe on top of an insulated column to produce very high electric potential. During the early stages, this type of accelerators faced operational difficulties in raising the potential beyond a certain limit. But at a later stage, such limit could be raised to much higher level when Van de Graaff vertical machine designed to generate as high as 10 MeV was constructed by Oxford University group with a plan to raise the final energy to 20 MeV of a complete electrostatic generator and integrate the total system as an injector stage for Rutherford High Energy Laboratory.

Cockcroft-Walton on the other side took a separate route and concentrated in developing a voltage multiplier circuit which could raise the low input voltage to higher level of output voltage. The duo ultimately succeeded in developing a two-stage voltage multiplier circuit which later extended to achieve higher potential for particle acceleration by increasing more number of stages. The Cockcroft-Walton model of multiplier circuit as it is known had two banks of series capacitors alternately connected by rectifiers acting as changeover switches according to the output polarity of the supply as shown in Fig. 1.1. It is sometimes necessary to use one or more stages of voltage doubling circuits when acceleration voltage is required to be raised. This type of multiplier circuits are still widely used injectors for higher energies. In that case, the no-load output voltage is twice the peak input voltage multiplied by the number of stages (N), where the number of stages is equal to the number of capacitors in series between the output and the ground.

Direct method of accelerating particles with sufficient energies had some inherent limitations due to nonavailability of suitable devices in developing accelerators of required energy. This deficiency encouraged exploring the possibilities of indirect acceleration, the basis of which is the time-varying aspect mentioned in the equation earlier. The search resulted in the development of two types of indirect accelerators, resonant and non-resonant types, depending on the mode of operation

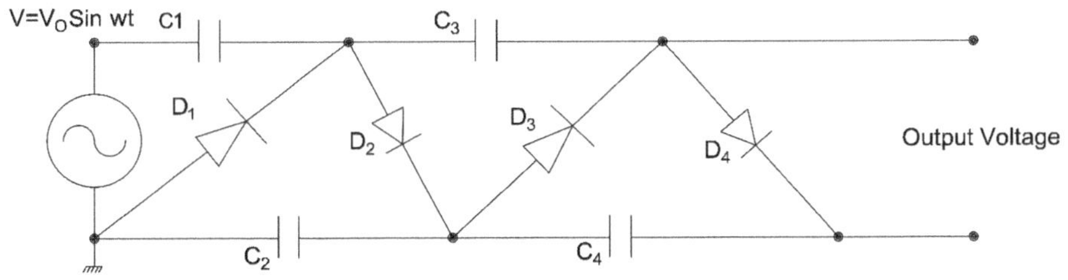

Fig. 1.1 Cockcroft-Walton voltage multiplier circuit

and the periodic changes of the electrical field. In one case, the principle involved with such functioning termed as linear accelerator where the particles travel in a straight line to arrive at gaps at the right phase of r.f supply or move in step with a travelling electromagnetic wave provided in the suitably design structure, whereas in other type particles travel a region of electric or magnetic field in approximately circular orbits for synchrotrons or traverse cyclically for cyclotrons and gain energy in each revolution. These two were the most prominent types of resonant accelerators. The principle of resonant type linear accelerator was attempted first by Wideroe (1928), connecting a number of drift tubes in series in increasing order of length to which an a.c voltage was supplied from a cyclic source, and the electric potential across these drift tubes was periodically reversed to continue the flow of electrical charges as shown in Fig. 1.2. This arrangement was made in such a way that alternate drift tubes were connected to the same terminal of an r.f generator. The polarity of the supply to the drift tubes ensured that, the acceleration of particles took place only on the forward direction and prevented the particles from being decelerated during anti-peak period by the reverse electric field. For effective functioning, the maximum oscillating voltage differential between electrodes needs to be injected at the right time when the particles cross each gap of the circuit. The length of the drift tube increases with the increase in velocity and required the adjustment of frequency. Thus, it became possible to introduce a new form of indirect acceleration with the linear acceleration of charged particles which moved between centers of successive gaps in one complete cycle of oscillation. In this case, the total energy output of the particle is represented by:

$$\mathbf{E} = \mathbf{N\,q\,V_p} \tag{1.4}$$

where E = energy output, V_p = peak voltage applied between the electrodes, q = elementary charged particles, N = number of accelerating electrodes.

It is important to note that higher the frequency of the acceleration voltage, the more would be the acceleration per length and shorter would be the accelerator.

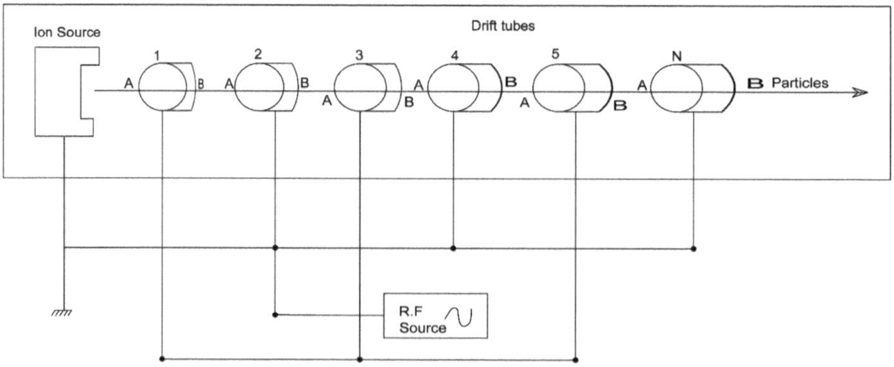

Fig. 1.2 Winrobe drift tubes in series

However, linear accelerator did not receive much attention until high power microwave oscillators made it popular after World War II. In this respect, the development of accelerators can broadly be divided into two phases, pre-World War and post-World War, and the in-between period termed as the intermediate phase was somewhat dormant or of low pitch for accelerator activities because most of the scientists and engineers were directly or indirectly involved in the preparatory war assignments for their respective countries. Van de Graaff, Cockcroft-Walton, Wideroe, and Lawrence were all pre-World War eminent scientists and engineers who contributed enormously in developing the initial particle accelerators.

In the techniques adopted by Van de Graaff and Cockcroft-Walton, the particles could be accelerated only once by the available electric field due to restrictions of voltage in generating higher energies whereas Wideroe and Lawrence took different routes and diverted their schemes of development from direct to indirect acceleration functioning in resonant mode. During the same period, both concentrated in the application of time-varying electric field through two distinct methods which distinguish in developing two different types of resonant accelerators, linear and circular types, respectively. Their efforts were to raise the accelerator energy level more than what could be achieved in the electrostatic type developed by other two. When Wideroe was busy in developing linear accelerator on one side, Ernest Lawrence concentrated in developing cyclic accelerator on the other side. In fact, indirect accelerators of protons in drift tubes used resonant cavities in which electrodes were introduced to direct the fields and enabled particles to be properly shielded from field reversals. This was found acceptable to the researchers because its advantages lied with the acceleration of light particles as well as heavy ions like mercury. The disadvantages in developing higher energies for accelerating light particles such as alpha particles required long vacuum tube and this invited practical problems in the construction of such tube. Since all three types of accelerators developed by Van de Graaff, Cockcroft-Walton, and Wideroe faced one or the other disadvantages like voltage limitations or length limitations in generating higher energies, the search continued to find viable alternatives free from such disadvantages. The phases of development of electrostatic accelerators, linear accelerators, and circular accelerators continued globally to achieve higher goals in raising the level of energies gradually with time. To sort out the earlier difficulties, the development of accelerators with time-varying field was found to be suitable in achieving higher energies and offer many advantages. Ernest Lawrence of the University of California expounded the principle that the motion of an accelerated particle perpendicular to a magnetic field results in the particle being taken a curved path. The theory presented was mainly directed towards developing cyclic accelerators which opened up new possibilities. This explanation of Lawrence changed the future course of research activities and initiated to implement the available theory into practice by bending the particles under the influence of a magnetic field and allowing the particles to move through the same electrodes repeatedly to gain higher energies. The result of such findings ultimately made a significant jump in the conception and realization of cyclic accelerators. One such especially designed cyclic accelerator over electrostatic and linear accelerator is termed as cyclotron with orbits consisting of a series of arcs of circles of discrete but of increasing radii in a constant magnetic field, whereas the other model termed synchrotron accelerator designed with orbits of approximately constant radius with a changing magnetic field. Cyclotron, the first cyclical accelerator, invented by Lawrence to establish his theory, adopted the simplest form of design with two semicircular hollow D-shaped sheet metal electrodes called "Dees" placed face to face inside a vacuum chamber between the poles of a fixed field magnet where a static magnetic field is applied perpendicular to the electrode plane to which r.f voltage is connected across the narrow accelerating gap between the "Dee's" as shown in Fig. 1.3. When protons or other positive ions are injected at the center of the magnet between the electrodes, the magnetic field deflects the path of the particles inside the "Dee's" and accelerates through the electric field twice per revolution as they spiral outward from the center of the machine. The particles are accelerated through electric field only and not by magnetic field which deflects the particles to take a semicircular path inside the "Dees." Again, by

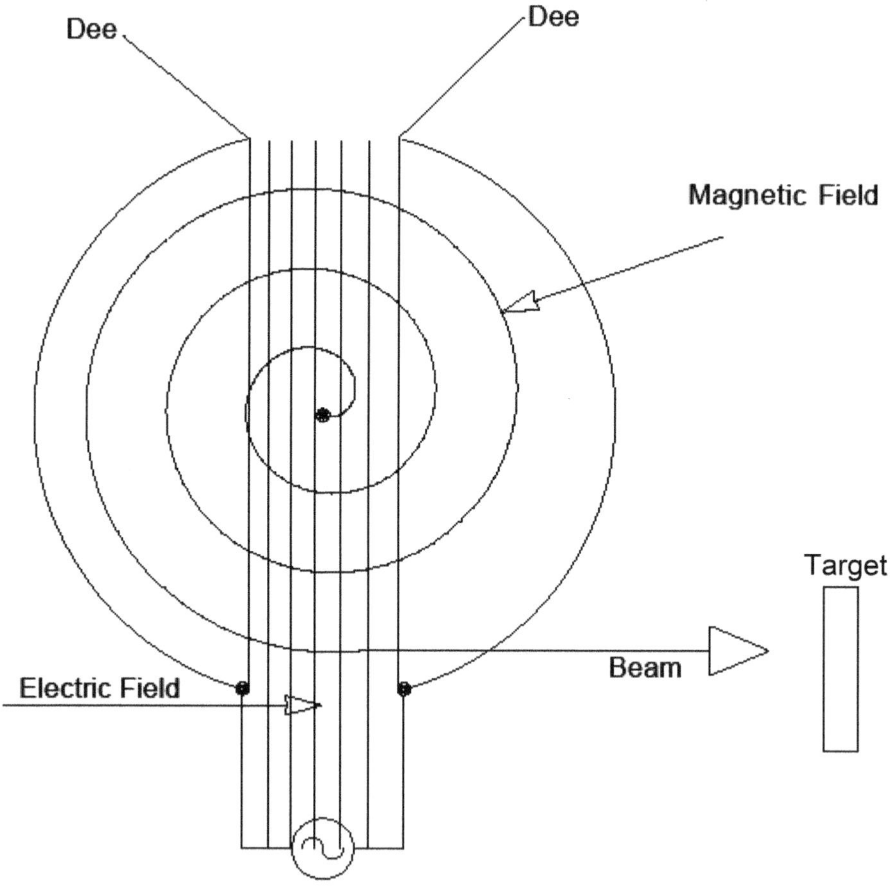

Fig. 1.3 Schematic of cyclotron structure

studying the simple relationship between the forces acting on the particle Prof. Lawrence realized that the increase in the radius of the path taken by the particle is compensated by the increased velocity of the particle if the magnetic field, the charge of the particle and the particle mass remains constant. The relationship between mass m, charge q, field B, and the frequency of the r.f source is given by:

$$f = \frac{Bq}{2\pi m} \tag{1.5}$$

Cyclotrons cannot accelerate negative and neutral particles and the final energy of the particle accelerated by the r.f voltage is dependent on the strength of the magnetic field and the diameter of the accelerating chamber. Actually, there are a number of differences in the operation of cyclotrons and synchrotrons which ultimately favor for synchrotron (Terwilliger n.d.). For example,

- Cyclotrons accelerate particles along an outward spiral path and are held in that path by a static electromagnetic field perpendicular to spiral path.
- In a cyclotron, conducting 'Dee's shield the particles from the electric field when they are away from the acceleration gaps, whereas, in synchrotron the electric field is confined to a small region by a radio-frequency cavity.
- The magnetic field in cyclotron is kept constant with time so that the particles need to spiral out to larger radius as they gain energy but in synchrotron the magnetic field is increased in synchronism with the accelerated field.
- In cyclotron, the particles are accelerated through electric field only and not by magnetic field which deflects the particles to take a semicircular path inside the "Dees." Again, as long as the protons are nonrelativistic, the particle velocity and radius both go up together. Also, in uniform magnetic field, for nonrelativistic particles the revolution period $\mathbf{T_R} = \frac{2\pi m}{Bq}$ does not depend on energy.
- In a synchrotron, since the radius is kept constant, the revolution frequency and the acceleration voltage and frequency must increase directly with the particle velocity.

There are even other differences between cyclotron and synchrotron. It is interesting to note that the concept of developing powerful accelerators utilize the basic principles of electromagnetic theory where the second term of Lorentz force is

$$\vec{F} = q\vec{E} + q\left(\vec{v} \times \vec{B}\right) \tag{1.6}$$

i.e., q v B is equated to the centrifugal force (mv^2/R). Hence, in order to maintain a circular path of radius R of accelerated particles unchanged, the magnetic flux density B must increase as the energy of the particle increases. Since, the magnetic flux density B, which is proportional to the energizing current I of the magnet system, it becomes one of the most determining factor of the proton synchrotron to maintain the constant radius for the accelerated particle. The activities of synchrotron

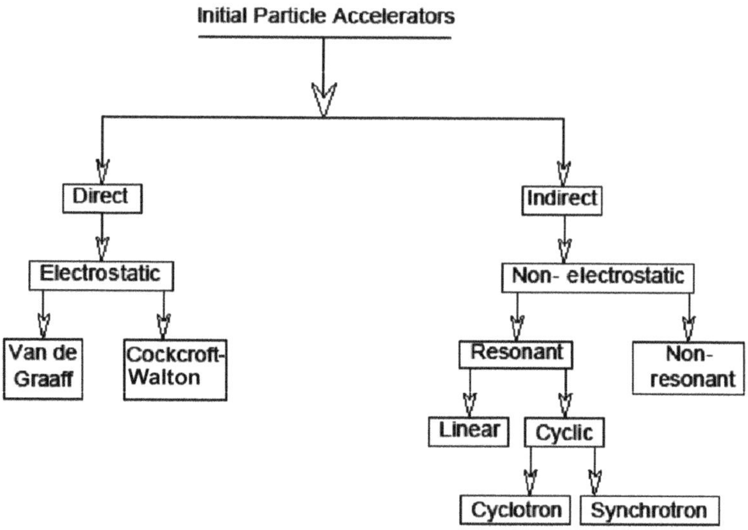

Fig. 1.4 Initial particle accelerators

accelerator with its advantages over cyclotron continued with the development of available technology support to fulfill the demand of the researchers. With the progress in the development of particle accelerator activities from early 1950s, a new chapter with distinct resonant types of synchrotron accelerators as shown in Fig. 1.4 begins with many new achievements to contribute to science, technology, industrial applications, medicines, and ultimate welfare of mankind.

References

Bryant PJ (n.d.) A brief history and review of accelerators. CERN, Geneva
Terwilliger KM (n.d.) "High energy particle accelerators"

Chapter 2
Power Supplies for High Energy Particle Accelerators of 1950s

Introduction

"Synchrotrons," the most powerful tool to fulfill the demand of the researchers to work with higher particle energies from 1950s started a new era, a new beginning after a long waiting period. Since then, the on-going research and development activities in various accelerators were continuing on different areas, to name a few; raising of energy levels to disintegrate nuclei with much higher energies in colliders, providing user-oriented beam dynamics, finding power supply alternatives for changing pattern of magnet excitations to reduce operation cost, changes in the generation of r.f power, upgradation of rectifier devices, changes in the converter topologies, adoption of new strategies for control systems, and many other developments for improving performance. Separate solutions were searched out to handle critical factors that restricted improved performance of the most powerful accelerators. All the hurdles were given due importance and resolved slowly after the end of World War II, when the scientists and technologists came back to take up their left over jobs with new vigor and enthusiasm to find improved ways in generating higher energies. Their activities continued accordingly to reach the set goals on above-mentioned research activities with the enormous practical knowledge and experience gained from the field work of war time assignments on projects related to atomic energy, radar, radio-frequency transmission, converters, control system, and many other related areas. At present, there are a number of high energy particle accelerators around the world but the most prominent and talked about high energy particle accelerators providing as much high energy as possible on particle physics are LHC and Tevatron. After the development of Large Hadron Collider at CERN, Geneva, enormous excitement was generated among the scientific community and the general public, to know the latest result of the experiment being conducted in that accelerator where, the physicists from all over the world were engaged in the

P. K. Dey, *Power Supplies for Synchrotron Accelerators*, Springer Tracts in Electrical and Electronics Engineering,
https://doi.org/10.1007/978-981-97-7606-1_2

fundamental research to find the latest result on "God's Particle." A Large Hadron Collider is designed to accelerate two beams of protons to energy level of 6.5 TeV each and cause them to collide head on, creating center of mass energy of 13 TeV. The Tevatron also a powerful collider accelerator at Fermilab, Batavia, Illinois, where the structure of the system is very similar to LHC, could accelerate particles from 200 MeV of Linac output to booster and with storage ring synchrotron assisting accelerator facilities to raise the energy of the protons and anti-protons to 980 GeV in opposite directions to collide at 1.96 TeV. LHC and Tevatron are actually developed as special type of accelerator complexes where, low energy from linear accelerator is amplified in booster synchrotron and storage ring arranged in cascade to generate the highest energy for final acceleration. Similar accelerator complexes are also developed in other countries as well to continue their research program with higher energies. On fourth of July, 2012, when CERN unveiled the mystery and announced the discovery of the ultimate particle, scientists and technologists were jubilant and celebrated the success of the experiment for achieving the long cherished goal. It was indeed an achievement and a proud moment for the scientists and technologists because this success:

(i) Pushed forward the frontiers of knowledge one step further and enhanced the determination to achieve the highest peak.
(ii) It accepted new technologies for the accelerators and field tested the advanced type of detector facilities particularly developed for this experiment.
(iii) It showed how global participation, involvement, fund sharing, and teamwork of scientists and technologists could help in establishing the most difficult theoretical stipulations through dedicated experiments.

This enthusiasm very well reminded me of the similar situation experienced in the early 1960s at the Rutherford High Energy laboratory (RHEL) at Chilton, Berkshire, England over the development of 7 GeV proton synchrotron. It was also a technological wonder and marvel to me when I visited the place for the first time in a tour program organized by the Institution of Electrical Engineers, London in 1964. Actually, LHC developed at CERN having 27 km of circumference is of no doubt a technological wonder and included the best of engineering and technology to build up such an accelerator but, other synchrotrons of 1950s such as 1 MeV of Birmingham University, 3.2 MeV of Brookhaven National Lab, 6.2 MeV of Lawrence Lab, University of California, 7 GeV of RHEL machine, and CERN's 25 GeV proton synchrotron were also considered equally path breaking. Actually, when the accelerator projects were coming up in early 50 s, it was realized that, to obtain energies of high magnitude, very large particle accelerators are required. But, the allocated fund to meet such exorbitant cost of the project was beyond the scope of any individual European country to venture in realizing the need for research with higher energies. In order to come out from such financial constraints and meet the consistent demand and eagerness from scientists to continue with the research programs, it was ultimately decided to join hands and attempted for a collaboration which made history. This initial effort with keen interest to forge the scientists and technologists resulted in the development of CERN's accelerator complex. Thus, the European organization for nuclear research (CERN) was formed with a group of 11 member states in 1954 to provide joint funding facilities for research in high energy particle physics at Geneva. This resulted at the first stage in the development of a large proton synchrotron of 25 GeV which came into operation in 1959 and created the record of being the first particle accelerator with the highest energy operating during late 50 s. It received the feather of being the most remarkable and established as a pioneer in the field of accelerators. All these synchrotrons were prominent and marvelous in design more than half a century back with differences in the energy level of the particles, in the selection of equipment, different in sizes and ratings, but similar in their system arrangements. The operational experience of these accelerators contributed largely in the advancement of initial technology and subsequently showed the torch for adopting in other synchrotrons. Even today, 26 GeV CERN's synchrotron forms a key link to CERN's 300 GeV Super Proton Synchrotron (SPS) which came into operation in 1977 6 years after the member states signed the agreement to develop the accelerator complex for higher energies and it is also a wonder in itself. The size and technology involved in building the proton synchrotrons including Rutherford Laboratory's Nimrod may look miniscule today in comparison to the giant size and the application of latest technology at LHC at Geneva, but it was an accepted fact that NIMROD, Cosmotron, Bevatron, 25 GeV CERN's proton synchrotron, and 300 GeV SPS all adopted the best of available technology of the time. The experience from the past has been shared to steer the course of action for the future in a direction which is expected to give positive result.

At the outset, I must confess that my second visit to RHEL was to join as an engineer in 1966. I am indeed grateful and indebted to my seniors, Director Dr.T.G. Pickavance, Chief Engineer Mr. Percy Bowles, and the Principal Scientific Officer Mr. M.J. Marchbanks at that time in RHEL, who provided me with necessary guidance to adjust in the uncommon territories and took keen interest to send me frequently to attend specialized lecture courses at Harwell Post Graduate Education Centre to familiarize with the latest state of the art in technology and upgrade theoretical knowledge in the specialized field from the experts from universities, research institutions, and reputed electrical industries on various subjects. It is therefore expected that the experience of Rutherford High Energy Laboratory which is embedded in my system will certainly be shared through specific references (Bowels et al. 1963; Fox et al. 1963; Rohtiget et al. 1963) and highlighted here more often in the description than other accelerators.

It was interesting to me for the first time, looking at "Nimrod," the huge machine was that the breaking of the smallest particle requires the maximum of energy with the total setup of huge electrical power equipment and the best of available engineering skill. The most striking difference with the manufacturing, production, or process industries was that there one could see the produced material and quantify it but here such feelings were absent because the nature of the final product was different and not visible. The development of very high kinetic energy of the particle was required for bombarding at the set target and thereafter recording and evaluating the structure of the disintegrated nuclei to find the result of the collision by the experimenters to attain at some conclusion was the objective of the facility. But, how charged particles were to be produced, how higher energies could be generated, how such kinetic energy could be imparted to the charged particles, how it could be guided to the target with maximum of beam intensity, how the result of the disintegrated particles after the bombardment could be detected, evaluated, interpreted, and how all such requirements could be practically implemented were serious questions which required a total team work and close bonding of scientific and technical activities.

Basic Physics and Process Flow Sheet of Protons

The involvement of huge cost for scientific experiments needs to be result oriented to serve for the better interest of mankind in the larger areas. Accordingly, the advancement of particle accelerators is focused on the following research areas with some particular interests:

- High energy particle accelerators.
- Ion sources/cancer therapy.
- Neutron sources.
- Light sources.

In order to get some immediate benefits, stress is given on the prominent and specific application fields which opened up new areas of research such as:

(a) Scientific applications, including high energy physics, environmental studies and application in medical diagnostics, and other areas.
(b) Industrial applications, chemical, mechanical including blasting, ion implantation in the semiconductor industries, plasmas, and modification of surface properties of materials, etc.
(c) Synchrotron radiation is being used in a variety of applications such as preservation of foods and food processing. Synchrotron light sources find applications for basic research and technical developments in varied fields like atomic and cluster physics, condensed matter physics, chemistry, material sciences, and crystallography, and it is probably in medical sciences that accelerators have found wide field of applications such as isotope production, diagnostics, therapy, and treatments.
(d) Opened up vast application field in the development of superconductor and its application to magnet systems and other areas.
(e) Opened up new concepts for magnet excitations.
(f) Opened up new technology for high power switching.
(g) Opened up new concept in the design of r.f power supplies.
(h) Opened up new techniques of converter topologies and its consequences in power supply system, new technique for ripple reductions.

(i) Opened up new schemes of VAr generating systems to reduce voltage fluctuations affecting the neighboring customers and test the effective tolerance of the national grid supply to highly fluctuating load.
(j) Opened up new field for developing sophisticated detectors.
(k) Opened up new methods for digital control, modeling, and computing techniques.
(l) Opened up new instruments with improved accuracies for current measurements and develop new type of flux measuring equipment.

Application of synchrotrons in Fig. 2.1a and classification of synchrotron accelerators in terms of energy storage bearing characteristic features of some distinct type shown in Fig. 2.1b are important particularly, for identifying their operation. Imparting highest kinetic energy to the charged particle being complex, the required energy level is raised systematically through several stages as mentioned below:

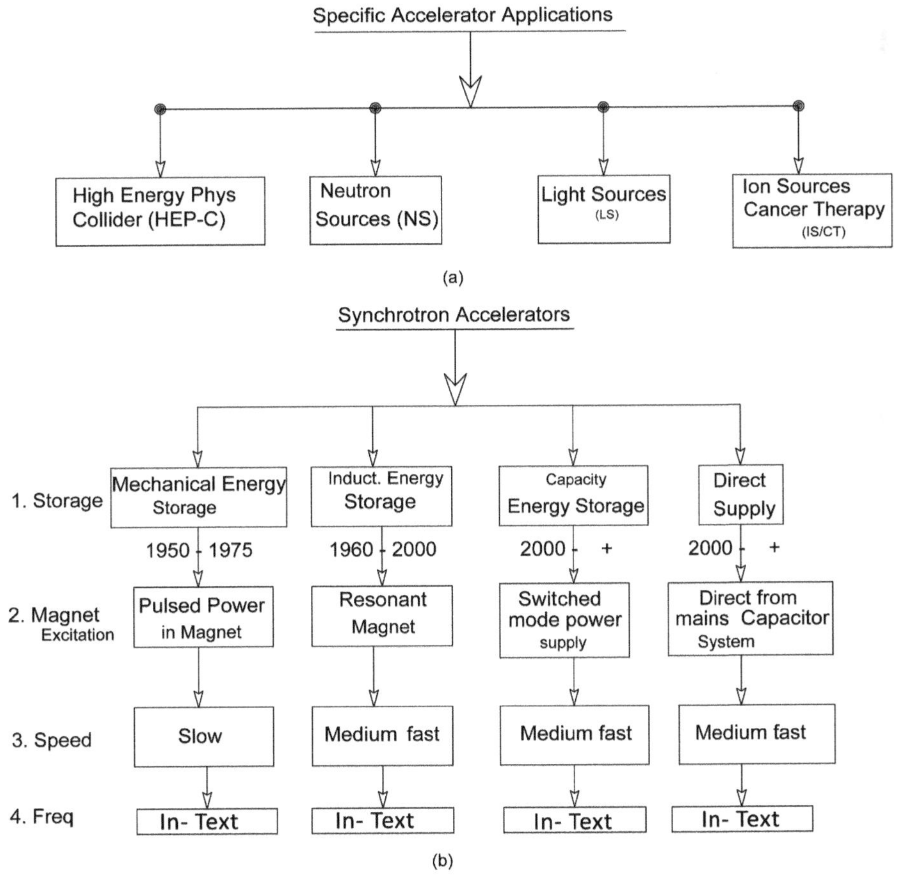

Fig. 2.1 (**a**) Application of synchrotron accelerators. (**b**). Classification of synchrotron accelerators in terms of energy storage

1. First stage is the production of charged particles for which there are two different sources:
   (i) In one case, protons are generated in an ion source and the voltage is raised in Electrostatic Generator before feeding the charged particles for linear acceleration.
   (ii) The other source uses electron gun for the generation of electrons by hot cathode. It begins the process when high voltage is supplied to a heated cathode to produce pulses of electrons for the linear accelerator. In this case, the voltage between filament and cathode is stabilized in order to hold the temperature of the cathode constant.
2. Second stage is to impart initial acceleration to the particles through linear accelerator.
3. Third stage is the injector part, where the accelerated charged particles are injected into the desired position of the proton synchrotron with required initial energy.
4. Fourth stage is the circular accelerator or the synchrotron, where the centrifugal force of the particle is equated to the Lorentz force to maintain the circular path of radius R unchanged. Briefly, the equations given below are important to understand the operation of the synchrotron accelerator.

$$\mathbf{mv^2 / R = evB \ldots;} \tag{2.1}$$

$$\mathbf{R = mv / eB} \tag{2.2}$$

$$\mathbf{f = \frac{v}{2\pi R} = \frac{\mathit{eB}}{2\pi \mathit{m}}} \tag{2.3}$$

$$\mathbf{m = m_0 / \sqrt{1} - \left(\frac{\mathit{v}}{\mathit{c}}\right)^2}$$

where R = Radius of the circular path; B = Magnetic flux density; f = frequency of accelerating voltage; C = velocity of light

$$\mathbf{e = 1.6 \times 10^{-19}\,As; \qquad m_0 = 1.67 \times 10^{24}\,g;}$$

Since these accelerate the protons nearly to the speed of light, their behavior can no longer be described by classical mechanics, but requires the help of the theory of relativity according to which, with increasing speed, an increasing proportion of the input energy is converted to inert mass. In order to keep the radius constant, eq. (2.2), the flux density B in the air gap must always be proportional to the product m × v. The flux density, in other words, the energizing current I of the magnet system which is proportional to B becomes the determining factor in the design of the proton synchrotron. But, the task of controlling the flux density to fulfill the set conditions invites some other difficult technical problems.

5. Fifth stage is guiding the beam through beam line magnets before extracting and bombarding at the target.
6. Preparation of the target area; to detect the result of the disintegration after collision and record, evaluate, and interpret the data to establish the findings of the experiment.

In proton linear accelerator, the particles develop their initial kinetic energy as per the design of the PLA and inject this energy into the torus of the proton synchrotron where very high vacuum is maintained. The particles are then accelerated further during each revolution till it reaches the maximum energy. At the time of acceleration in synchrotron, some important operational conditions are as follows:

A) The criteria of minimum initial energy of the protons at the entry to the torus of the magnet and before the start of acceleration must be maintained to keep the particles on their radial orbit as the circulating protons are particularly sensitive to magnetic irregularities at relatively low velocity.
B) Particles are not allowed to attain their full acceleration immediately after injection and therefore, the voltage necessary to raise the acceleration to maximum has to be imparted slowly in the form of voltage "kicks" during each revolution, in other way, the particles should not pass through the accelerating voltage at once.
C) Particles are allowed to make rotation several times in succession within a very short period to develop the energy from the potential field.

With the advancement of time, how booster and storage ring synchrotron arrangements are connected in tandem to offer higher energy facilities is shown in the figure. In fact, it can be simply viewed that, it is customary to accelerate the beam to the initial energy level first in linear accelerator, then to the desired energy level in booster synchrotron and thereafter, transfer the beam to the storage synchrotron. The function of the storage ring is to store beams of high energy particles to be used as and when required depending on the design of the facilities offered.

Engineering Design Considerations

I must confess at the beginning that although I was not connected with the Nimrod from the inception of the project, but I fully appreciated the relentless effort of the scientists and engineers who made this project successful. 7 GeV Nimrod, proton synchrotron was engineering marvel from all aspects such as civil, mechanical, electrical, instrumentation, and control. The involvement of civil and mechanical has been briefly described to impart the cardinal points only with the exception of the electrical engineering part with instrumentation and control which has been taken up separately to emphasize the subject "power supplies for synchrotron accelerator."

(a) Civil Engineering—Rhel.
Civil engineering along with other engineering services during project handling played a significant part in the civil design and foundation of Nimrod along with associated equipment. The primary constraint of the civil design was the safety of personnel and equipment. When the synchrotron would be operational, the protection from intense radiation produced would be of serious concern and therefore, the civil design was to ensure that the injector and magnet rooms were adequately shielded to provide necessary safety in all respects. The effect on the temperature of the surroundings due to heat generated from various machine components in an enclosed area and adequate shielding for the protection of instruments from background radiation was given due importance in the design. In view of that, semi-ground construction was adopted for the main building and both concrete and earth were employed for shielding. Another impressive design was the concrete tunnel leading an approach road to the machine room with hundreds of cables stacked systematically on both sides of the tunnel. The atmospheric temperature of the underground tunnel was kept constant by a very efficient air-conditioning plant in order to exclude any possibility of geometrical change due to thermal expansion.
(b) Mechanical Engineering.
Design and manufacturing of magnet sector plates was a highly skilled job and without going too much into mechanical detail it can be mentioned that the manufacturing of electromagnet for 7 GeV proton synchrotron was another engineering marvel. The ring-shaped 7000 ton electromagnet had 160 ft. diameter and consisted of equal core shaped sectors and disposed in a circle comprising eight octants. The magnet sectors were fabricated at the Wolverthon works of M/s Sankay of England and for ease of transportation from the manufacturer's works site to the project site of RHEL, at Chilton, Berkshire, it was developed in pieces of 336 sectors. Each magnet sector was built of 45 silicon steel plates, 10′5″ high and 9′3″ wide. With regard to high standard of specification demanded, manufacturing of magnet plates and its alignment played a special roll. Sufficient precautions were taken during manufacturing process to meet the highest tolerance requirement specified for each activity with regard to annealing, shearing, drilling, and machining at the factory. A special procedure was evolved to ensure that the surface finishes of the plates were of highest standards. The proper installation and positioning of these magnet sectors at the project site and their alignments, apart from all other activities had to undergo stringent inspection and scrutiny to satisfy the challenging technical demand. The geometric alignment of the magnet of the proton synchrotron presented unusual problems at the time of installation and commissioning because of accuracy required to maintain the plane of rotation during commencement and completion of bursts of pulses; otherwise, the accuracy would have been affected due to sinking or tilting of the magnet foundation.

(c) Electrical Engineering.
The selection of power supply for the magnet was important for the operation of synchrotron accelerator for which extensive studies and many different combinations were evaluated before accepting the final design. System design for power supplies for RHEL and 25 GeV CERN was more or less similar in line with other previously designed synchrotron accelerators. Nimrod produced a high speed beam of protons at an energy level of 7 GeV which was bombarded on targets arranged as per the requirement of the experiments. Resultant nuclear reactions were then studied to provide data on the behavior of elementary particles. In addition to Nimrod with its demand of currents up to 10,500 A at 15 KV, other equipment such as Proton Linear Accelerator as initial energy source and r.f. generator designed to give the required voltage kick to boost up the energy level during each rotation of the particles, made a heavy demand on the power supply equipment. Energies of about 15 MeV from the Proton Linear Accelerator (PLA) were injected into a large evacuated torus of the synchrotron maintaining a vacuum of 10^{-7} Torr. These particles were further accelerated by r.f. generator in a cavity to impart energies at regular intervals inside the magnet. NIMROD was designed to produce pulses of 7 GeV protons at a rate of about 28 pulses/min with about 10^{12} protons in each pulse, this is equivalent to a mean proton current of about 0.1 micro amp, whereas the 25 GeV proton synchrotron of CERN was designed for maximum flux density of 1.2–1.4 wb/m^2, acceleration cycle of 12–20 per/min and anticipated protons about 5×10^{10} in each pulse. The limit to the energy of the protons produced by a synchrotron accelerator is set mainly by the radius of the ring which is 100 m for CERN and the maximum magnetic field of 1.4 wb/m^2 for bending the protons. Since NI MROD required pulses of 15 MeV up to 0.002 s long duration at 1 s intervals, the energy from the injector had to be pulsed.

The r.f. accelerating unit imparted 5 Kev to 7 Kev energy kicks to the protons on each revolution and after a period of 0.7 s, by which time the protons would make almost one million revolutions to reach the desired 7 GeV energy. Power consumption to the accelerator cavity was of the order of 1 MW. During each revolution when the energy of the protons were increasing, the magnetic field had to increase in proportion and in synchronism with the voltage kick to a peak value of 1.4 wb/m^2, in order to confine the protons into their circular orbit of radius R. When the protons reached the maximum energy, the beam was taken out with the help of extraction unit to strike in a target. After that, the magnetic field was gradually reduced to zero and kept in readiness for injection of a further pulse. The whole process was to be repeated again. The nature of the demand of the magnetic field during each pulse from the starting of current rise to current zero is shown in Fig. 2.2. So far, it has been briefly explained how the protons were injected to the torus of the synchrotron magnet and the requirements of varying magnetic field with the application of proportional voltage kicks to maintain the proton in its orbit of radius R. But, the complexities involved in the design and development of electrical system along with associated electrical circuit to follow the desired characteristics posed a difficult hurdle to the electrical engineers, and it needs to be highlighted to

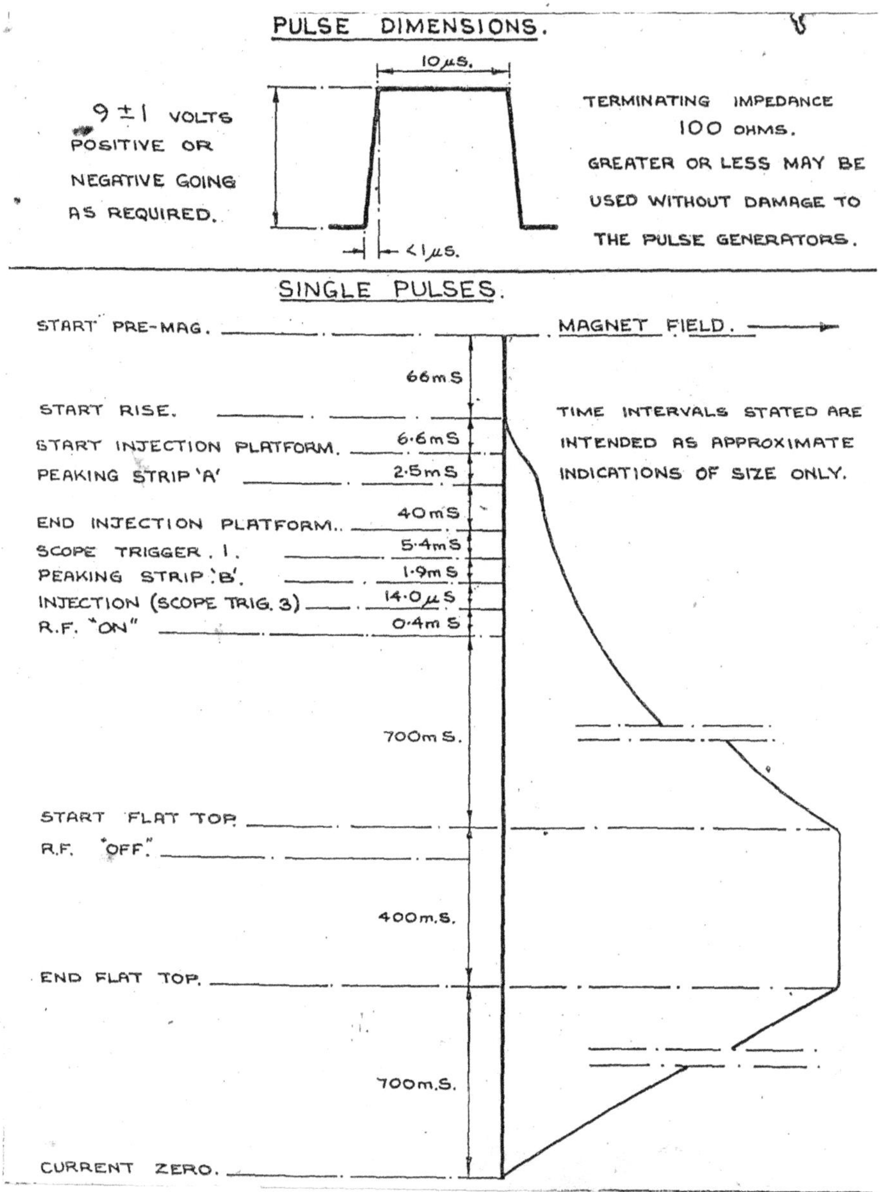

Fig. 2.2 Magnetic field characteristics of Nimrod—RHEL

understand the complexities and enormity of the power supply requirement for the magnet as well as for other critical equipment during that period. Present day engineers won't be able to appreciate this problem as most of the technologies adopted earlier to solve the hurdle have become obsolete, even the textbooks are not available for consultation and the courses are withdrawn from the syllabus long back. Although engineers engaged in similar jobs today may not find it interesting to know the problems faced earlier and the discussion on outdated technology may sound futile, they would definitely appreciate how the system approach, being the same, and the advanced technology can simplify the complex design. It is therefore thought worth to record the history to understand the technical challenges of the earlier period by highlighting the developments in stages and the hurdles faced by the engineers to reach at the solution with the available technology of the time.

The emphasis will be placed on the 7 GeV proton synchrotron—Nimrod of Rutherford High Energy Laboratory to establish that all other synchrotrons have passed through similar experiences. Most often comparisons will be made with CERN's 25 GeV proton synchrotron, which was operational during the same period to consider it as some reference (Ludwig et al. 1959). Actually, there cannot be any comparison between the two units as the difference in the design features of the two synchrotrons was prominent for various reasons. Firstly, the energy levels of the two units were different secondly, the difference of energies of injected proton beam thirdly, the CERN's design was an earlier model than RHEL's Nimrod fourthly, the difference in the selection of redundancy provision to maximize the operational efficiency of the synchrotron and thereby reduction in the loss of experimental time fifthly, separate converter configurations to contain the harmonics level and lastly, the most essential of all was the involvement of cost of installation during different period of times. It is only to be noted that the problem in both cases was almost same. The very simple reason the project was challenging to the electrical engineers was that the steep demand of current during energizing of the magnetic field within a very short period and afterwards recovering of the energy stored in the magnet coil and continue to repeat the program. It will be worth mentioning at this point that the effect of resistance and magnet time constant variation resulting from temperature changes were minimized by introducing water cooling the magnet windings when the magnet itself was air cooled by the air conditioning of the magnet hall. The two Figs. 2.3 and 2.4 bear three elements characteristics of the magnet demand. The first one shows the characteristic curve of Nimrod and the second one shows the characteristic curve of CERN synchrotron, respectively. Characteristic details of the two units are shown in Table 2.1.

Actually, the three element characteristics as shown in the figures mentioned are very common in variable speed drive applications such as reversing rolling mills, mine winder drives, and many other process applications, but the difference lies in the exacting demand of voltage and current during acceleration period and decaying period. The common features in both cases of characteristics are that the first part is the accelerating part in case of variable speed drive, whereas in case of synchrotron accelerator the magnetizing current is gradually rising; the actual time of current rise is depending on the value of the required magnetic field. The second part is the flat top part, or the steady-state part of the speed in case of variable speed drive, whereas, in case of synchrotron accelerator it is the period during which time the magnet current is held constant. This condition is obtained by arranging only half of the converters to operate as rectifiers while the remainder functions as inverters. The

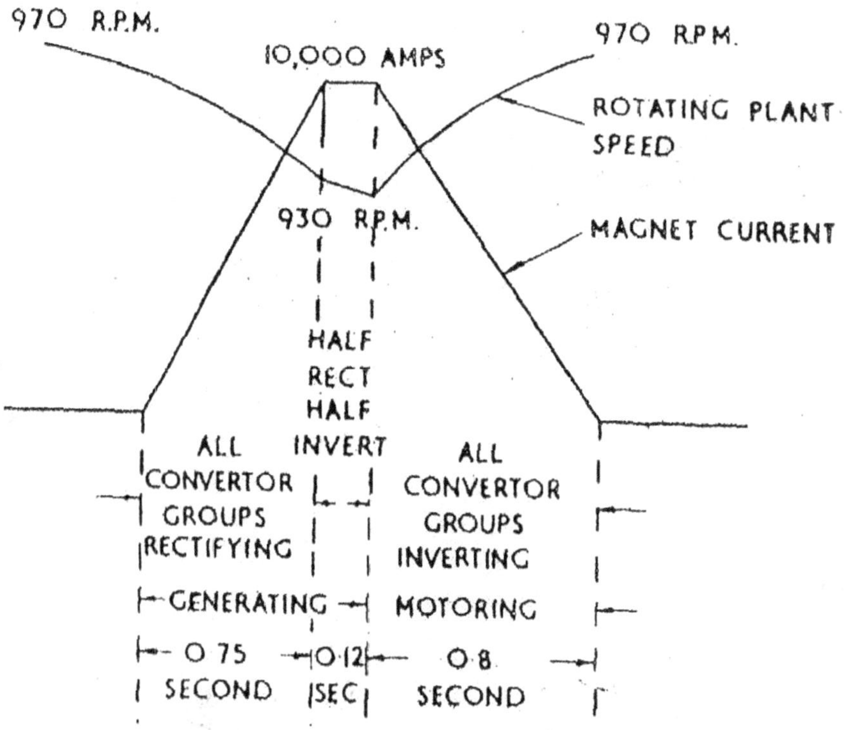

Fig. 2.3 Characteristic Curve of "Nimrod Power" With Permission from: Rutherford Laboratory

third part of the characteristic in case of variable speed drive is the decaying part and during this part braking is applied properly so that the speed of the machine is gradually reduced to halt at the required position and ready for the next cycle to start without any difficulty. But, in case of synchrotron accelerator, it is the magnet de-energizing part; during this part, the energy stored in the magnet coil is withdrawn and returned back to the supply. During this period, all the converters will be in inversion mode. As can be seen in the table, CERN 25 GeV synchrotron did not have flat top period. Interestingly, there were some confusion with such choice of allowing a flat top period, but in this regard characteristic of 1 GeV accelerator of Birmingham University illustrated a typical field density waveform and explained the necessity and the advantage for introducing the flat top strategy. The main advantage which accrues upon the introduction of flat top is that the extracted particle beam emerges slowly over a controlled period of time instead of short bursts (Barney 1967). Since Nimrod project was designed much later than the other accelerator projects and developed by British scientists, it may have followed the Birmingham strategy. In case of variable speed drive, the factors mentioned above can be varied according to the demand of the process, whereas such variation is restricted in synchrotron application. The critical demand is that the magnet must

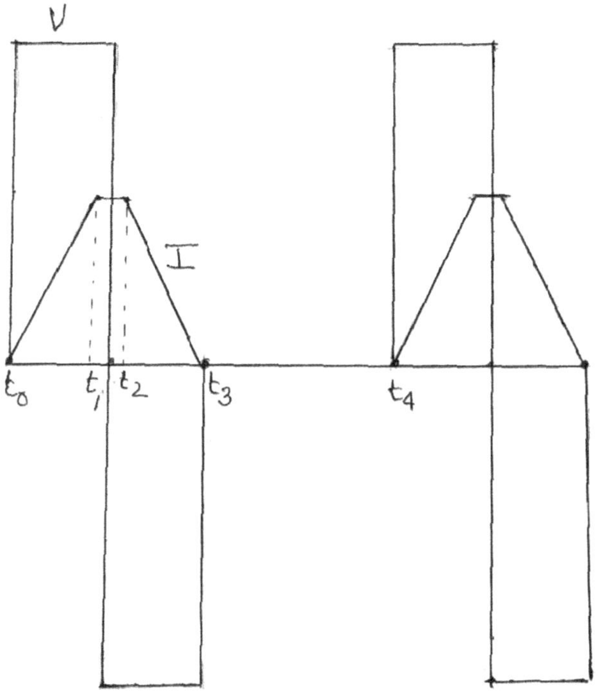

Fig. 2.4 Characteristic curve of CERN accelerator with permission from CERN

Table 2.1 Characteristic details of the two units

Project	KV	KA	Rise time (s)	Flat time	Decay time (s)	C/S	Speed fluctuation r.p.m
7 GeV Nimrod	11	10	0.75	0.12 s	0.8	28	970–930
25 GeV CERN	5.4	5.0	1.0	Very low- 0	1.0	12–20	3200–2800

follow the set reproducible program which exactly means that, after every pulse, the magnetic field has to come back to its initial state. The restoration of magnetic field to an initial definite state is essential after every pulse for synchrotron so as to perform a repetition rate of 28 cycles per second for Nimrod and 18 cycles/s for CERN's synchrotron. Of course, the repetition rate for accelerators also can be changed with the change in the reproducible program as per the requirement of experiment or the capability of the machine. The return to the initial state is governed by a factor known as remnants of the magnet material. A definite reproducible initial stage can be attained only when de-energization of the magnet always follow the same curve. It may be possible that the other advantage for adopting flat top strategy was to ensure following of such reproducible program strictly thereby, increasing the stability. It is also important for the voltage at the magnet really to be zero during the pause period in order not to affect the initial state due to slight change in flux and therefore subjected to greater tolerances; otherwise, the protons would depart from their prescribed orbit soon after injection. In fact, the magnet voltage also must remain constant during energization and maintain the reproducibility of the pulses within close limits. In Fig. 2.3 for Nimrod, the machines were acting as generators when the speed of the rotating plant falls from 970 rpm to 930 rpm, and during this period the converter groups are acting as rectifiers. Obviously, the CERN's requirement showed the flat top portion to be nearly zero, and it was almost triangular in shape. Since the magnet from the electrical point of view behaves almost as a pure inductance, the stipulated current rise is obtained automatically if the applied voltage is kept constant. Simplified pulse waveform of magnet current and voltage waveforms of Nimrod Magnet during operation is shown in Fig. 2.5 and at the same time a simplified characteristic of the alternator during operation is shown in Fig. 2.6 to emphasize the loading pattern. In case of Nimrod, the problem is to charge an inductor of 1.06 H with a maximum current of 10,000 A in a preset time of 0.75 s.

(i) After knowing the stringent demand of the magnetic field and the difficulties of pulsed power supply to the magnet, it may be assumed that possible solutions to the problems might have been looked into from three available choices:

1. Supply the magnet power from a d.c. source.
2. Utilize a.c to d.c. converter, i.e., Supply a.c. power directly to the rectifier unit and feed the d.c.to the magnet.
3. Supply the a.c. power to a motor–alternator set and then convert to d.c. through rectifier before feeding to the magnet.

In case of 1, in feeding the magnet from d.c. machines, the need for switching the magnet voltage on and off physically made this difficult. Earlier, for variable speed drive in industrial applications, Ward–Leonard drive was very popular, the application of which will be discussed later. Mechanical switching was ruled out because the frequent switching operations involved in pulse operation. It was only possible to connect or disconnect the d.c.voltage through the excitation of the machine, but the specified rise time of a few milliseconds would have been difficult to attain by this method. Therefore, supplying the magnet power from a d.c. source operating at shorter duration was ruled out except the Birmingham accelerator where such time factor was high.

The next possible method was to feed the magnet from converters directly connected to the supply network. In this case, the specified voltage rise would have been quite easily attained with grid controlled mercury–arc rectifier, but in that case, the supply network would have to possess a very large capacity to deliver the enormous power and withdrawal of stored energy from the coil during each pulse would create problem. It would also prove rather difficult to retain the specified constancy of the voltage and comply with the conditions imposed regarding allowable voltage fluctuation in the supply network. So, this method was also ruled out at that time.

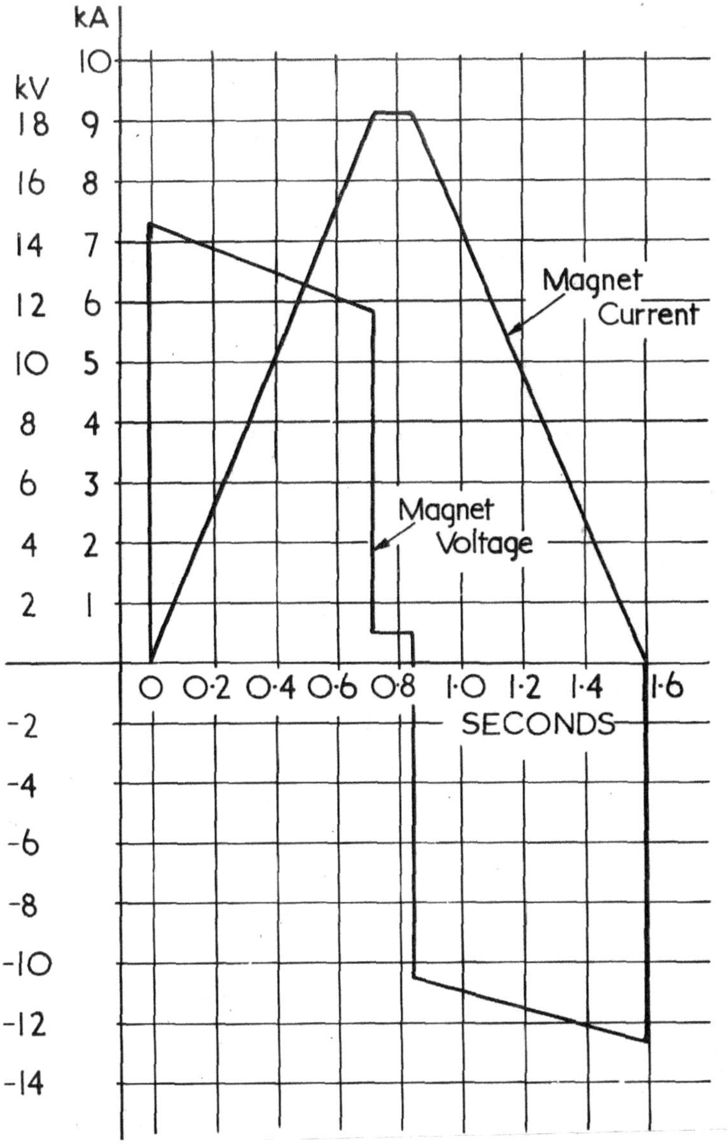

Fig. 2.5 Current and voltage waveforms for Nimrod

There remained therefore, the third method where the magnet was to be fed from a.c. motor–generator set followed by rectifiers. This arrangement should be able to comply with the requirements on the power engineering side, only the contribution of the fluctuating load on the supply network needed to be seriously considered before the final design. Since the magnet windings form a highly inductive circuit, a large amount of energy would be stored in the magnet coil after withdrawal of the

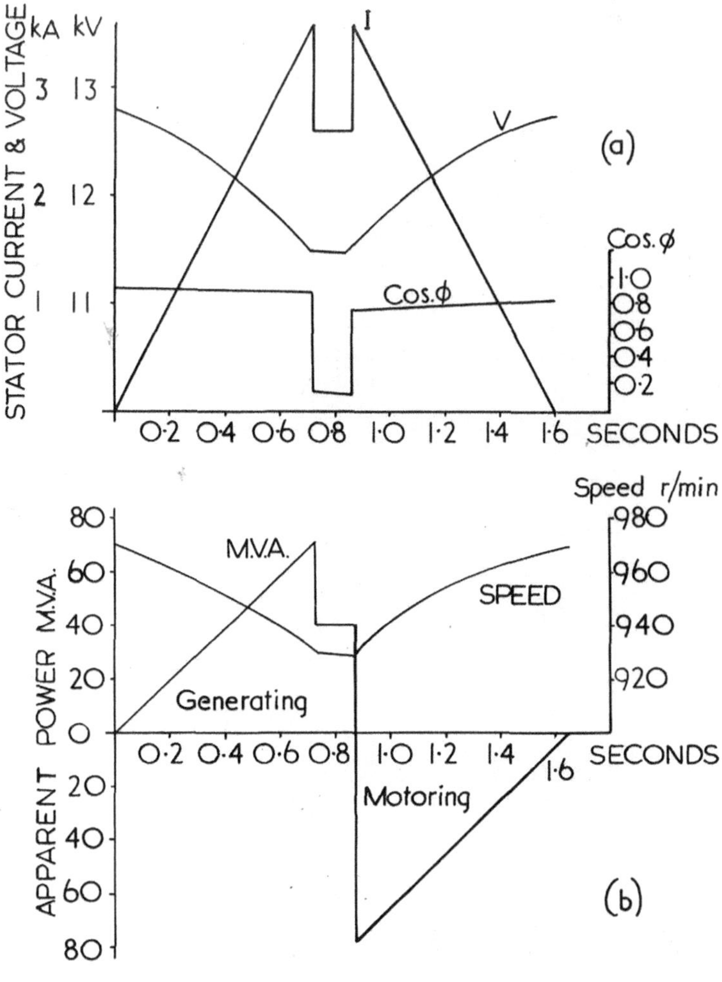

Fig. 2.6 Alternator loading for Nimrod

current. Hence, the withdrawal of energy from the coil and storing the energy in a technically feasible manner was important, and accordingly the application of flywheel in combination with the motor–generator was considered suitable. The addition of flywheel would definitely control the fluctuating load to some extent in the input supply. In such cases, harmonic generation originated from the rectifier circuit would be made acceptable to the supply authority through proper configuration of the rectifier network to limit the harmonics to the specified limit.

Such flywheel coupled to the motor generator set was common in industrial application as an energy storage device, even in some cases during emergency or interruption of power this scheme of stored energy is utilized as an uninterrupted source of supply for a very short duration. In synchrotron accelerator applications when peak and mean input power are compared, it will show how extraordinarily severe the conditions are for the power supply network to tolerate such fluctuating load. Such demand of fluctuating load can only be reduced with the help of energy storage units. The system must be such that, it not only be able to draw energy from the storage source during magnetizing part but, be able to accept the return of energy when the magnets are de-energized. The storage unit must therefore be so designed that it supplies as well as accept the return of large amounts of energy from the magnet while the input power from the supply network will be drawn only to meet the losses. Therefore, to meet the pulsating demand for the accelerator application, the combination of a flywheel coupled to motor and generator set in the same shaft was adopted. Incorporation of flywheel set to absorb or deliver energy mostly depends on the total mass and the diameter of the flywheel. The violent cycling in speed and load amounting to a typical fluctuation in speed from 970 rpm to 930 rpm in case of Nimrod and for CERN's installation, the speed variation was restricted to 3200 and 2800 rpm. Again, the speed of the synchronous machine could not be allowed to fluctuate unduly; otherwise, it will raise complications in control. The energies which could be obtained from the rotating masses are proportional to WD^2, where the diameter is selected from the difference between the squares of the maximum and minimum speed variation. Therefore, the larger the speed difference, the smaller the flywheel. In order that equipment is kept to a minimum and that the least demand is made on the incoming electricity supply a flywheel weighing 30 tons had been coupled on each motor–alternator rotor shaft for Nimrod, the amount of energy circulated in this way amounted to about 45 MJ. Each alternator was rated at 60 MVA at 12.8 Kv and weighed a total of 400 tons. The detailed ratings of the electrical equipment in each organization have been shown in Table 2.2. In variable speed drive for rolling mill, mine winders and in many other processes, any one of the three methods outlined above are selected according to the requirement of application. But the most interesting point to be noted here is that, earlier in variable speed drive application Ward–Leonard drive was common and the mercury–arc rectifiers were avoided. In all such application rotating machine was the only choice. But in case of synchrotron, the magnitude of current and voltage to be handled required some serious analysis before the selection of a suitable design out of the three available methods. The actual constraint was that, any deviation in the supply characteristics will create a serious problem to hold the proton in its orbit and jeopardize the experiment. The consequences of such demand of pulsating load to the electricity supply network, in other words, restrictions imposed by the local supply authority on the fluctuating loads and the installed electrical equipment needed critical examination before the selection of the system. A group consisting of 96 mercury–arc rectifiers for twin set of Nimrod and 24 mercury–arc rectifiers for CERNs with a

Table 2.2 Specification of rotating equipment

Projects	Alternator	Motor	Flywheel
7 GeV Nimrod	60 MVA, 3 ph, 50 Hz, 11.5 KV, 1000 rpm	5000 HP	30 tons
25 GeV CERN	46 MVA, 3 ph, 50 Hz, 6.9 KV, 3000 rpm	2600 KW	N/A

huge rectifier plants with all its accessories and control systems operating in open space made a special impact on excellent mechanism of technology.

Choice of Rotating Equipment

The system design was basically same for both the organization and similar design was followed by all the accelerators at that time. The differences were mainly on the setup due to operating energy, on the selection of equipment, choice of electrical systems, providing maintenance support, cost factor, factor of safety of equipment, and some other factors. As mentioned, the two synchrotrons had their specifications different, including the injector energy input to the machines. Therefore, it was natural that the selection of equipment ratings, flywheel size, transformer connections, and the rectifier configurations would be different. The most prominent part was the selection of speed of the set, more clearly, one unit selected two-pole machine whereas the other unit selected six-pole machine. It may be seen that the diameters of the flywheels were also different in sizes. Since the diameter of the flywheel is dependent on the square of the difference of the speed variation during pulse period, the dominant factor is the tolerable variation in speed during operation. But, at the same time, during the selection, the importance had to be given to restrict wide variation of the speed; otherwise, it would have serious effect on the supply network. Nimrod was a very expensive machine and therefore, it was designed to operate at its maximum efficiency. From the design, it was clear that RHEL had given due importance on the redundancy of the power supply to the magnet. The power supply to the Nimrod proton synchrotron was effectively in two parts. This twin arrangement of motor–generator–flywheel set was independent and provided some stand by protection against any serious breakdown on any one of the rotating equipment. In case of breakdown in any one unit, the operation on the other half could be continued to maintain power supply at reduced load so that it could at least serve some experiments to continue. But, in case of CERN's power supply arrangement, redundancy factor was not taken into consideration, there could have been a number of reasons for such decision but few possible reasons might be:

(a) the cost factor because the cost of the dual set would have been exorbitant at that time,
(b) the factor of safety chosen in the selection of equipment for the rotating machines were kept much higher to give adequate cushion for any breakdown,
(c) it might be that the experimental load was not expected to be high enough for which the breakdown time could not be spared.

With the massive weight of the rotating mass of motor–alternator–flywheel set of about 400 tons and with the fluctuation of speed from 970 rpm to 930 rpm during each pulse, it was anticipated that an alarming problem will be created with regard to vibration during operation of the Nimrod synchrotron. In order to restrict such tremendous vibration to a limited value, adequate damping arrangements were made by mounting the total assembly on a reinforced and post-stressed concrete foundation block weighing about 1200 tons. Again, this foundation block was supported on 80 spring units and 12 viscous damper units. This foundation block ensured that virtually no vibration would transmit to nearby plant and buildings. Apart from the vibration damper arrangement mentioned, each set was provided with bearing vibration and shaft eccentricity measuring instruments. At this point however, it may be appropriate to establish how true and effective the choice of dual set was in the electrical system design for Nimrod when the operational history of the Proton Synchrotron at RHEL is looked into. Actually, the first motor–alternator–flywheel set was installed and commissioned in conjunction with the converter plant during the early part of 1962 and the power supply was used for Nimrod magnet until August 1962. The second set was installed during the latter half of 1962 and commissioning was completed during Feb 1963. The synchrotron accelerator continued its operation until a major fault occurred on number 1 alternator in 1965. As per the site investigation report, the fault was due to a crack developed in rotor pole end plates, indicating typical characteristics of fatigue failure. Since the decision to adopt two sets was made and commissioned, the failure of one alternator could not totally disrupt the operation of the synchrotron. The synchrotron was kept operating even during this breakdown period of one set at lower capacity. The investigative work in connection with the cause of failure by the engineering team along with the manufacturer of alternator took some time. End plates of forged steel of increased thickness were substituted for the original castings. The rotating set was subsequently restored back to circuit for operation to continue after some modification in the design and strengthening of protection system. It could be noticed that the power requirement for the CERN's unit was much lower than the Nimrod's. This is evident from the choice of motors, alternators, the flywheel set, the choice of transformers connected to the rectifiers and the rectifier configurations.

Transformer Connections and Selection of Rectifier Groups

The effects of harmonic distortion due to rectifier on a.c. supply impose constraints in the selection of transformer connections and rectifier configurations between the alternator supply and the connected load. At the time of designing and selecting of such a system due consideration was given to the rectifiers, the largest harmonic producing devices connected to the supply. But, it was not possible to lay down a generally applicable definite limit of permissible harmonic voltage distortion at the point of common coupling since the permissible value depended very much on the various items of plant and equipment connected to the supply system. However, the system design had to satisfy the strict stipulation regarding harmonics imposed by the supply authority. Any load of rapidly varying magnitude produces a corresponding variation in the supply voltage and because of this the electricity boards had to introduce limits of voltage variations which required being acceptable at the point of common coupling for fluctuating load. The back firing of Hg–arc rectifiers was also of considerable interest during operation due to inverse voltage immediately after arc extinction which influenced the rate of change of current. The operation

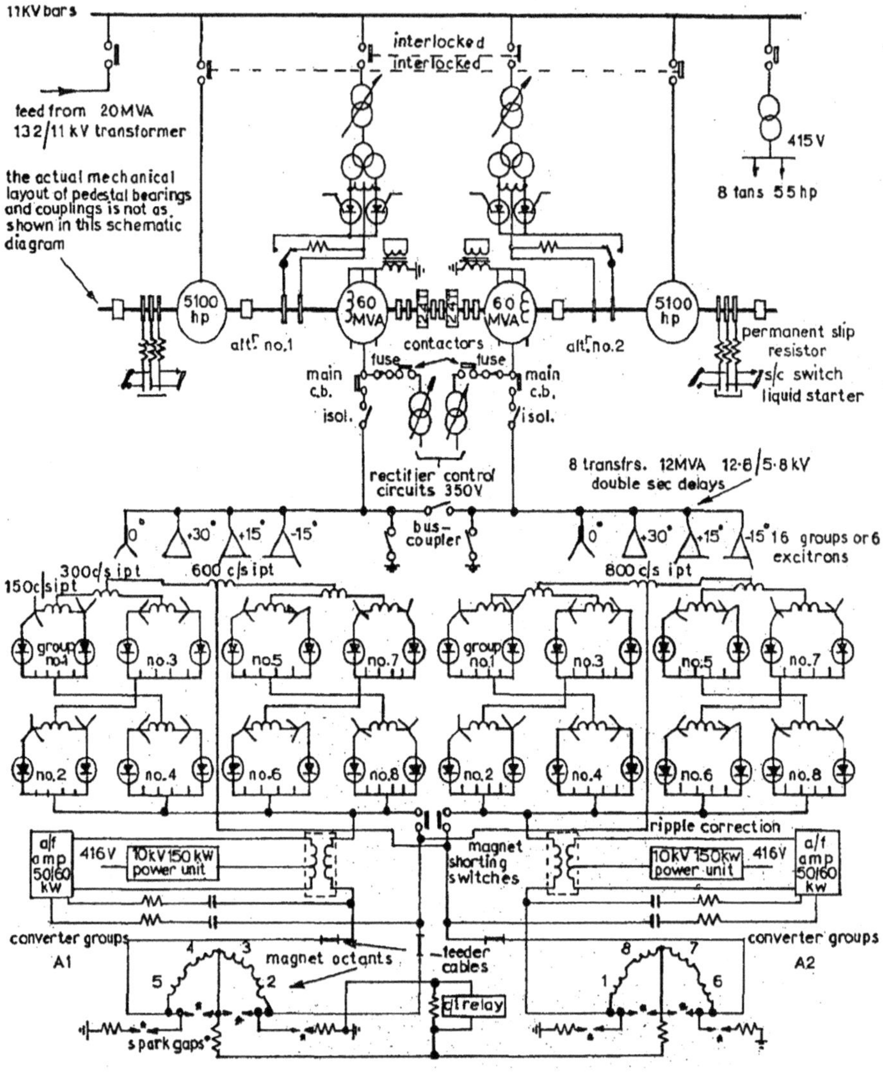

Fig. 2.7 Power plant schematic with transformer-rectifier Nimrod

and the problem of improving rectifier reverse voltage were found to be difficult both technically and economically. The rectifiers were designed as high current/low voltage units and the peak current in Nimrod operation was less than half of this high continuous current duty. Power plant schematics and the layout of the complete installation with the transformer and rectifier configuration adopted for Nimrod is shown in Fig. 2.7, and the basic features of rectifier connections for CERN synchrotron is shown in Fig. 2.8. As can be seen in Nimrod, each alternator fed to four rectifier transformers, each having a primary rating of 12 MVA. The primary

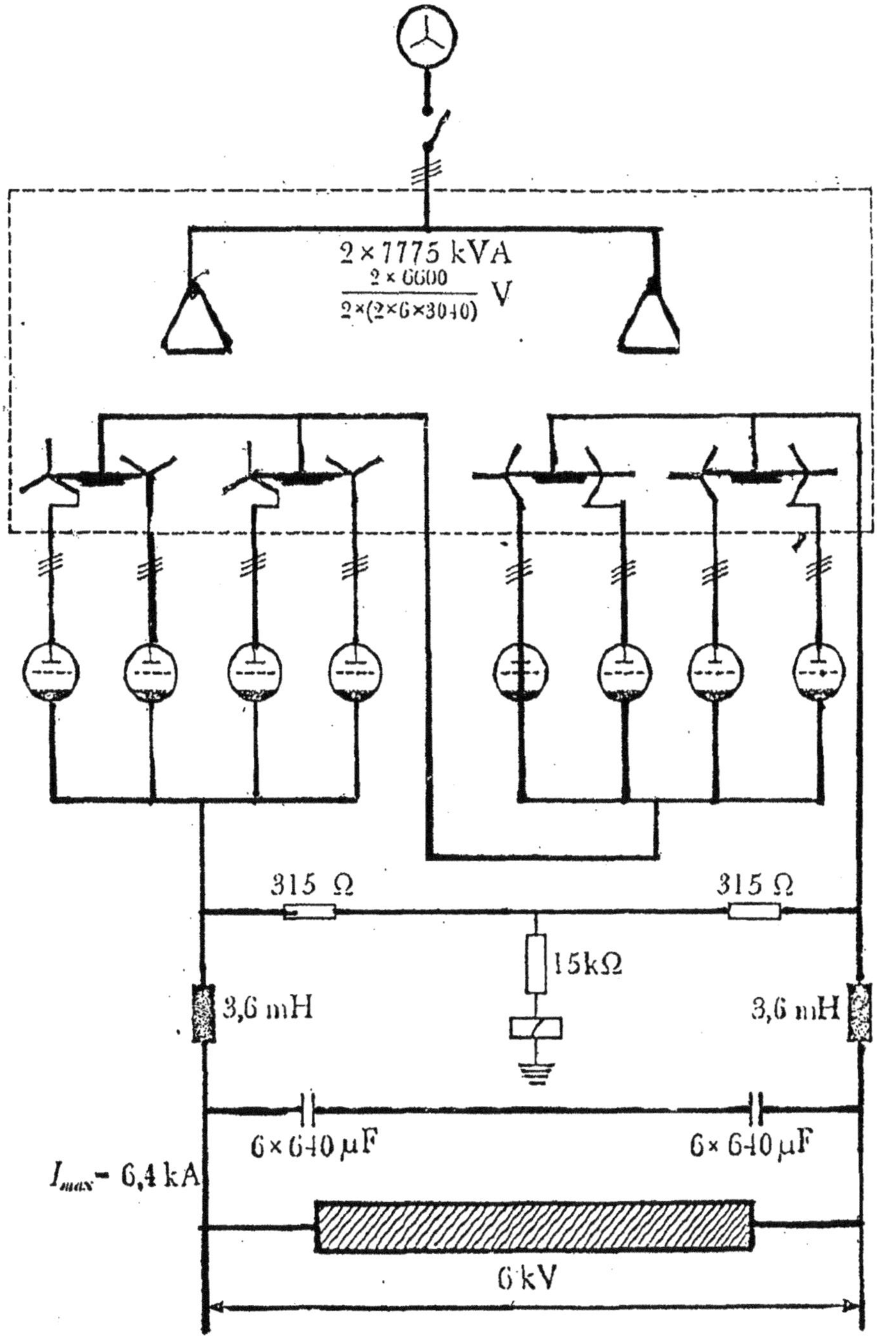

Fig. 2.8 Power plant schematic with transformer-rectifier-CERN/ Brown Boveri

windings of the four transformers were connected in star, delta, extended delta plus 15° and extended delta minus 15°degrees. Each transformer had two secondary windings connected double star with inter-phase transformer arranged as a 24-phase system. Also, each transformer secondary fed six single anode steel tank mercury–arc converters of the water cooled type and arranged in groups. Eight groups with six converters in each group feeding to the four octants, i.e., 48 rectifier loads were connected to each motor–alternator set. Similarly, the other motor–alternator set supplied power to remaining four octants which resulted in total ($8 \times 6 \times 2 = 96$) converters. The reason for such dual connection was that if any serious defect developed on one half of the power supply, the other half could still be used to pulse the magnet. In case of CERN, two transformers were used in the primary side and the primary windings of the two transformers were connected in polygon, the secondary's being mutually displaced by 30 electrical degrees, so that output voltage of the rectifiers had a 12-phase ripple. Each transformer had two secondary windings in a double 3-phase connection with an inter-phase transformer for a d.c. voltage of 3000 Volts and a peak current of 6400 Amps. In this case, two groups having four sets in each group were arranged with three rectifiers in each set ($3 \times 4 \times 2 = 24$). The outputs of the two units were connected in series to obtain the required voltage of 6.0 KV. The required d.c. voltage was produced by two series groups of converter units, comprised of 24 single anode steel tank mercury–arc rectifier. The converters were arranged in this way to reduce the harmonics in the d.c.voltage to minimum. In this case also water cooled single anode mercury–arc rectifier tanks were used to obtain a temperature distribution as nearly uniform as possible. Six tanks formed a set and each set being mounted on its own framework insulated from earth with the necessary accessories for firing and excitation. With regard to the choice of rectifier units, both the units had no other alternative but to choose single anode grid controlled mercury–arc rectifier because semiconductor rectifiers came much later. With the advent of semiconductor diodes and commercial availability of controlled rectifiers (Thyristors) from early 1960s with its various advanced forms, the design philosophy did changed. At the earlier stages of semiconductor rectifiers, some drawbacks were there in their application due to the limitation of power ratings, voltage ratings, current ratings, duration of turn-on, turn-off times, pulse firing circuits, etc. Thus, wider application of semiconductor devices was restricted but from late 60 s with the upgradation of power handling capabilities, improvements in dv/dt, di/dt characteristics, introduction of I^2t fast-acting fuses for protection, different combinations of circuit configuration with thyristor and diodes, the application of semiconductor devices for power control was universally accepted as best. Later on, the application area of the semiconductor diodes and controlled rectifiers with its various forms started dominating the total rectifier operation. The converter and inverter grade thyristors of much better form and improved version of semiconductor devices is now readily available for high power applications. Only interesting feature that had to be noticed in the design was the selection of the phase displacement of the transformers and the number of phase grouping of mercury–arc rectifiers. Actually, present-day engineers won't be able to appreciate this problem as the technology used earlier became obsolete from early 70s. The practicing engineers may now find the solution to be simpler because of the easy availability of the power handling devices, different forms of pulse firing circuits, simpler method of device cooling arrangements, suitable protective devices, electronic control system for monitoring of various parameters, acceptable technology for reducing harmonics, and suitable electronic control system for monitoring of other parameters .

Electrical Supply to Supporting Equipment

The support of the specifically designed power supplies to the following critical equipment and systems was important in the functioning of Nimrod as well as in all synchrotron operations.

(i) r.f power supply,
(ii) beam line magnets,
(iii) extraction mechanism and bubble chamber,
(iv) vacuum system.

R.F Power Supply

The protons were injected in the torus of Nimrod at an energy level of 15 MeV with an orbital frequency of 355 Kc/s and as the protons were accelerated to the energy of 7 GeV, the frequency was raised on a relativistic curve to 2.2 Mc/s, when the corresponding velocity of the charged particles reached very nearly that of light. It can simply be mentioned that the r.f structure had three main sections and the r.f field was in synchronism with the proton orbital frequency.

(a) the high power r.f. equipment providing the accelerating voltage,
(b) the low power r.f. equipment or primary frequency generator (P.F.G.), providing an accurate and controllable frequency source for the high power r.f. system,
(c) the beam control system, which provides a means of comparing the radial position and phase of the proton bunches with suitable reference value and of correcting any deviation via the P.F.G.

The complexities involved in the design of R.F power will be further discussed at a later stage.

Beam Line Magnets

High energy beam from particle accelerator Nimrod were guided to various experimental areas by beam lines using quadrupole and bending magnets. The total power required for the 14 beam line magnets for Nimrod installation was high, nearly to the order of 10 MW to steer the beam to the experimental area. Each beam line was supplied from separate stabilized d.c. source and these power supplies were developed either in-house or bought-out items from outside sources as per the required specifications. This surprisingly high power requirement was comparable to the enormous power drawn for the production of the proton beam termed to be the cost of beam transportation to the target. Stabilized current supplies to these beam line magnets were essential because any variation beyond the limit of 0.1% would result in improper steering and loss of particles.

Extraction Mechanism and Bubble Chamber

To carry out the experiments, it is necessary to extract the protons and send them up to the experimental areas for which a large pulsed magnetic field is applied to deflect the protons from their normal orbit into the extraction channel. Thereafter, detect and record the event of scattering after collision, collect data from counters, spark chambers for detecting and recording purposes, and utilize them for the analysis. It was one of the important phases of synchrotron operation where the beam of particles after acquiring the highest energy were directed towards the target set for collisions. In the collision of the extracted proton from Nimrod and the target, nuclei secondary particles were produced, these secondary particles were steered by the beam line magnets to the electrostatic separators where the so-called bullet was selected and separated. RHEL used bubble chamber to get result of such experiment. The bubble chamber recorded an "event" after an interaction between the bullet selected by the separator and the liquids in the bubble chamber. For extraction mechanism, sophisticated and massive equipment were required to separate and transport the particle beam to the target. The particles then can be detected in a number of ways, e.g.,

(i) by the scintillation of some materials when a particle passes through,
(ii) by the ionization produced in gasses when a charged particle passes through,
(iii) by the production of tiny bubbles when a charged goes through a liquid which is just on the point of boiling.
The momentum of a charged particle can best be determined by plotting its path through a magnetic field. To obtain the precision required for high energy particles, high magnetic fields over large volumes are required. Different liquids were used to investigate different effects. For Nimrod at RHEL, a magnetic field of 12 KG was produced by the liquid hydrogen bubble chamber magnets which made very heavy power demand. The power supply to the bubble chamber magnets was from 4 × 1 MW, 250 V d.c. generator sets and each generator set was driven by 1440 HP, 3.3 KV, 3 ph, 50 Hz, 740 r.p.m. squirrel cage induction motor. In Cosmotron however, it allowed the extraction of the particle beam for experiments located outside the accelerator. For most experiments, either the protons or beams of secondary particles are directed on to a target and the objective of the experimenter is to identify the directions of

particles coming from the target, their energies, their velocities, and if they are unstable their lifetime and what they disintegrate into to find the laws determining their production and interaction. In this regard, it will be worth mentioning that 300 GeV SPS which came into operation much later confronted with many problems in the field of engineering and technology while carrying out the experiments. The decision with regard to development of CERN's 300 GeV super proton synchrotron project will be discussed elsewhere.

Vacuum System

A synchrotron had a large doughnut-shaped vacuum chamber which was a tank usually made of stainless steel of suitable dimension to form the inner cell in which the particles were accelerated. Proper vacuum was of prime importance in the functioning of synchrotron because its objective was to minimize collisions between proton and gas molecules by reducing the pressure in the inner chamber to 10^{-6} mmHg. In order to create such vacuum, a large number of vacuum pumps were installed in each synchrotron. For CERN's 25 GeV synchrotron, near about 66 numbers of vacuum pumps were connected to the annular chamber which reduced the residual pressure to less than 10^{-5} mm of mercury. The vacuum system for 300 GeV SPS demanded high standard of engineering because there are nearly 4000 flange joints in more than 10 km of vacuum chamber for which the reliability is required to be very high.

Summing Up

The power system played an important part in the operation of high energy particle accelerators, the system design as well as the maintenance standards were kept high and strictly followed. Routine checking of technical parameters of all critical equipment laid down by the manufacturers and the technical cell reduced the breakdown to minimum. In order to give the maximum machine availability to the researchers, the teamwork and the co-ordination was of the highest order. In such projects, time was an essence and since it had high application potential, it was ensured that no bottleneck should come on the way to development otherwise, advanced research program would always trail behind others. Also, here lied greater scope for research and development in the superconducting magnet systems which would open up a vast application field. In order to achieve this goal, superconducting materials capable of sustaining high current densities in magnetic fields were considered for potential application for high energy physics experiments. It has been possible now to achieve high current densities in superconducting magnet systems which have been utilized in Large Hadron Collider to attain such incredible energy.

References

Barney GC (1967) New magnet power supply of the Birmingham proton synchrotron. Proc. IEE 114(9)

Bowels P et al (1963) Magnet power supply for the 7Gev proton synchrotron Nimrod, Pt-I. Proc IEE 110:561–572

Fox JA et al (1963) Magnet power supply for the 7 Gev proton synchrotron nimrod, Pt-II, rotating machines. Proc IEE 110

Ludwig EH et al (1959) The supply system for the magnet of the Proton Synchrotron of the CERN, Geneva. Brown Boveri Rev.; Vol,46 (6):327–349

Rohtiget K et al (1963) Magnet power supply for the 7Gev proton synchrotron Nimrod, Pt-III; mercury arc converters. Proc. I.E.E. 110

Chapter 3
Comparison of Power Supply Equipment of Synchrotrons of 1950s

Introduction

Synchrotron accelerators were considered as the major development projects for scientific experiments with higher energies from 1950s. In fact, it is a continuation of the earlier chapter specially prepared to throw some further light in the selection of design and developed as per the demand of strategic equipment before finalization of the system. At the initial stage, the scheme of power supply for magnet excitation of all high energy particle accelerators followed almost the same general structure. The realization of electrical equipment chosen for accelerator can best be explained and summarized through two fronts as follows:

(i) to project how critical was the design of power supplies,
(ii) how parameter deviations made the operation for all high energy particle accelerators difficult to provide reasonable performance without proven maintenance support.
Attempts have been made to justify the selection of equipment in general terms keeping aside the influence of magnet dynamics and without involving much on theoretical aspects. Actually, it all started soon after World War-II, when scientists and technologists took up their left over jobs with new perspectives and enthusiasms in developing 1 GeV accelerator at Birmingham, the first of its kind to focus on the technology of raising higher energies. The operational discrepancies gained from such project helped to identify the criteria for assessing the selection of equipment for subsequent projects. This Birmingham machine gave a good ground for understanding the complexity and the nature of the design requiring modifications, particularly specifying the advantage of introducing a gap of short duration after the magnetic field reaches its peak. Understanding the enormity of the task and the positive effect of a steady-state period, this scheme was included in magnet excitation for RHEL to improve

P. K. Dey, *Power Supplies for Synchrotron Accelerators*, Springer Tracts in Electrical and Electronics Engineering,
https://doi.org/10.1007/978-981-97-7606-1_3

Table 3.1 Synchrotron accelerators of 1950s in order of energy level

Sr. no	Institute	Location	Commissioning date	Energy (GeV)	Mean radius of magnet (M)	Peak power (MVA)	Storage capacity (10^7 J)
1.	Birmingham	UK	1951	1.0	4.5	10.2	7
2.	Cosmotron	USA	1952	3.0	11.3	26.0	12
3.	Bevatron	USA	1954	6.3	19.0	100.0	80
4.	Nimrod	UK	1961	7.0	18.75	108.0	39
5.	CERN	Switzerland	1959	25.0	100.00	141.0	148.0

beam dynamics but not tried in CERN. It is a subject of lengthy discussion but tries to quantify and illustrate the extensive lateral cooperation of designers, specialist consultants of the projects, manufacturers of each selected components required to share the problems in overcoming technical limitations and sort out complex demand of synchrotrons. Comparison of power supply equipment of synchrotrons revealed how difficult was the exercise in selecting the type of electrical rotating equipment, transformers, and rectifiers to strengthen the importance of design. This comparison of selective equipment for different accelerators is all the more necessary for future engineers while targeting to improve performance. It was an accepted fact that whatever might have been the technical difficulties, working in such environment and frequent confrontation to the unusual problems gave a separate inspiration to the engineers. It has been decided to choose only five initial synchrotrons to illustrate the critical points raised during the selection of electrical equipment considered to be pioneer for comparison and analysis after arranging the accelerators in order of energy level from minimum to maximum as shown in Table 3.1.

This analysis is made to find similarities, linearities, and discrepancies of the accelerators to establish how important were the bonding between technologists of different projects. Although, any such comparison of accelerators is unjust and uneven because, the nature of the accelerated particle, the energy level to be provided, difference in the required intensity of the particle beam, the required pulse period, the pulse per minute to be handled, the nature of the experiment to be conducted, the facilities to be provided to the experimenters, the mean radius of the synchrotron to be selected, the availability of funds for the particular synchrotron project which dominated the selection and other important parameters such as magnetic fields, frequencies, energy at injection, and intensities were altogether different for each of the project and shown in Table 3.2.

Even then, ignoring all that deviations, it tried to emphasize how exchange of vital technical information from experts and the cooperation received from manufacturers and other contractors helped in executing ultimately what was best for the projects. The comparison of the equipment is made to highlight the design choices of the past and emphasize that the sharing of new ideas and cooperation is all the more necessary before the project design to minimize the hurdles and avoid pitfalls

Table 3.2 Comparable parameters of synchrotrons of 1950s

Sr. no	Synchrotrons	Mag. Field initial (gauss)	Mag. Field final (gauss)	Freq. at injection (KHz)	Freq. final (MHz)	Energy at injection (MeV)	Intensity
1.	Birmingham	220.0	12,500.0	330.0	9.3	0.5	3×10^9
2.	Cosmotron	311.0	14,000	400.0	4.2	3.6	2×10^{10}
3.	Bevatron	300.0	16,000	360.0	2.5	10.0	2×10^{10}
4.	Nimrod	296.0	14,000	354.6	8.04	15.0	1×10^{12}
5.	CERN	150.0	14,000	290.0	9.55	50.0	3×10^9

in operation. A view of the massive twin motor–alternator set combined with flywheels for Nimrod (RHEL) is shown in Fig. 3.1 to highlight the general scheme of arrangements of rotating equipment for other accelerators.

Motors

The primary consideration in the selection of drive motor for synchrotron accelerator duty was to draw minimum power required during operation. This was possible because the initial power demand was only to start up and compensate for the running losses in rotating machineries, winding and I^2R losses in motor, generator, and in magnet coils which consumed significant power together with substantial losses in transformers and converter operations. The voltage drop due to arcing of Hg–arc rectifiers or ignitrons during operation was also very high for which elaborate cooling arrangements were installed for their operation.

Table 3.3 which compares the rotating equipment of the synchrotrons shows that the motor ratings were lower for all the synchrotrons as expected. Interestingly, since the energy of the Bevatron accelerator (6.2 MeV) (Green, 1953) was double that of Cosmotron 3.0 MeV (Kresser, 1952), the motor rating of Bevatron was 3500 hp., 8 pole wound rotor motor just double that of Cosmotron with 1750 hp. which was expected as the losses incurred by Bevatron was double that of Cosmotron. A linear relationship in the selection of the motor ratings was drawn because in this case the other parameters were found to be just double than the other one. However, such linear relationship of Cosmotron and Bevatron is an exaggeration and cannot be extended to other synchrotrons. Proportionate relations of energy in GeV and hp. of the motor are not expected such as in the selection of 5100 hp. motor for 7 GeV RHEL-Nimrod and 25 GeV CERN synchrotron. Actually, correct ratings for the motors depended on the permissible power swing, the speed drop, and the fact that in each pulsing cycle the motor must average out the entire losses of the plant. If a normal design was used, a ratio of 1.5 to 1 for peak/r.m.s torque was considered suitable for the duty. In that case, the rating required for Nimrod should have been 4410 hp., but this was increased to 5100 hp. to provide a margin for various larger pulses without sacrificing the pulse repetition rate to an undesirable extent. Also, the

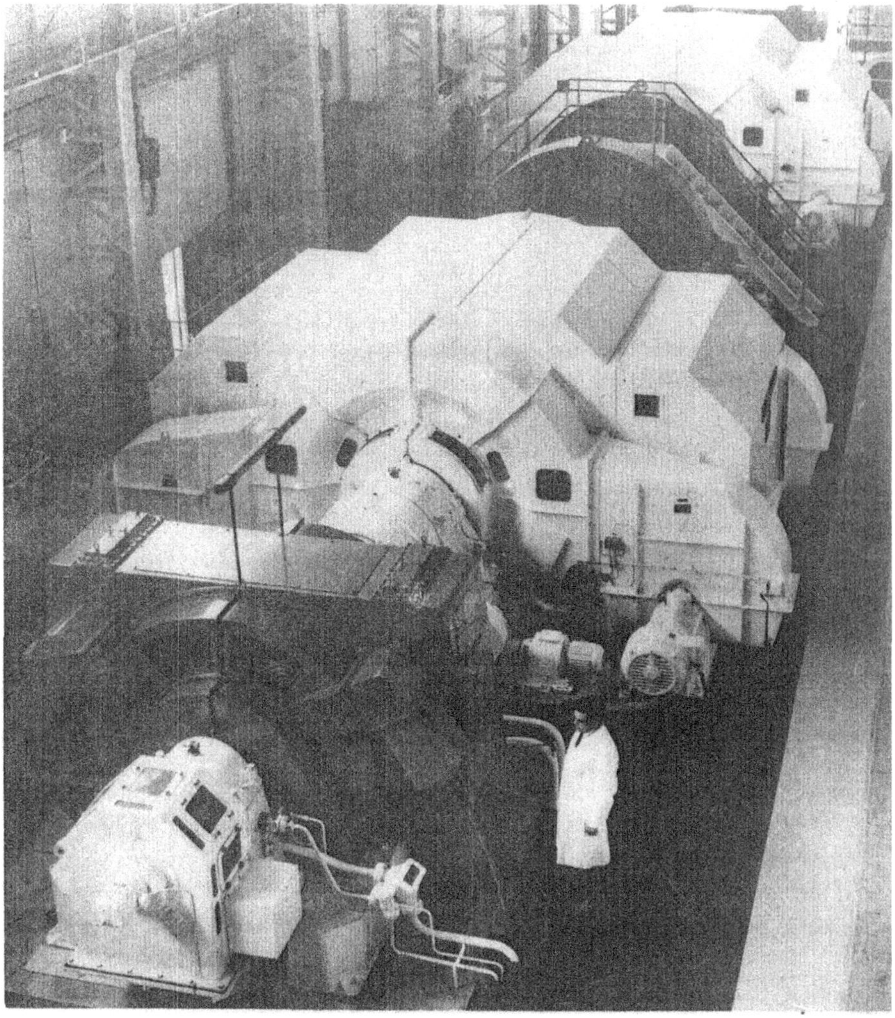

Fig. 3.1 Massive twin power alternator set for Nimrod

Table 3.3 Comparison of rotating equipment of 1950s

Sr. no	Synchrotrons	Motor rating (HP)	Gen. Rating (MVA)	Gen. Voltage (KV)	Peak current (A)	Speed variation In %	Flywheel weight (ton)
1	Birmingham	1500	20.0	N/A	12,500	N/A	36
2	Cosmotron	1750	20.0	4.2	7000	6	40
3	Bevatron	3600	46.0	3.2	8333	7	67
4	Nimrod	5000	60.0 MW	11.5	9150	4.1	30
5	CERN	3800	46.0	6.9	12,800	7.2	50

additional 1500 hp. compared with 3500 hp. of Bevatron to meet the difference of 1 GeV energy level between the two was only to cover the additional losses incurred in RHEL Nimrod operation due to its size of the alternator, flywheel, system configuration, and the availability of the nearest standard motor size from the manufacturer. The magnet power supply for CERN's 25 GeV accelerator was only one quarter of that of Nimrod and used a system allowing for a pulse without large power variations. On the other side, power supply for Birmingham proton synchrotron was different and being the first of its kind used d.c. generator for the magnet supply. The Birmingham synchrotron was developed to accelerate protons to 1 GeV for which M/G set comprised of 1500 hp. induction motor and D.C generator supplied field excitations to two parallel fields of the main generators mounted on a common shaft with 36 tons flywheel. The main magnet supply to the synchrotron was switched on and off through contactors. This form of configuration is very well accepted in industrial applications to supply emergency power for a short duration to the loads such as cranes or converters handling hot metals to bring the system to safe and secured positions to avoid any hazards due to sudden supply failure. The D.C machines had high losses and inherent maintenance problems apart from additional losses incurred in commutators and brushes. Such system was adopted by introducing a safety period, a long gap between the ends of the cycle period to the beginning of the next cycle which was important in this scheme of arrangements. As mentioned in the previous chapter, this scheme was not considered earlier, and this type of excitation system was later shifted to rectifier devices.

Alternators

The alternators were subjected to typical pulse loading and the distribution of the time periods shown in Table 3.4 are for different synchrotrons.

The comparison between the peak and mean input power gives some indication how extraordinarily severe the conditions were for selecting alternator suitable for a particular synchrotron. In selecting a suitable alternator, the factors such as the pulse rise time, pulse repetition period, amount of stored energy, and magnet losses with other electrical specifications like rotor type, number of poles, etc. played a

Table 3.4 Comparison of time periods of different synchrotrons of 1950s

Sr. no	Synchrotrons	Energization Period (s)	Flat top period (s)	De-energization period (S)	Idle period (s)	Total time (S)	Pulse per min
1	Birmingham	1.0	0	1.0	8.0	10.0	6
2	Cosmotron	1.0	0	1.0	3.0	5.0	12
3	Bevatron	1.75	0	1.75	1.6	5.1	11.7
4	Nimrod	0.72	0.115	0.80	0.665	2.3	26
5	CERN	1.0	0	1.0	1.0	3.0	20

significant role. If the rise time was made long to lower the peak power, the magnet I^2R losses would be large and the stored energy recovered would be low. Again, a short rise time would increase the peak value required to excite the magnet which in turn would increase the cost of the power supply. Synchrotron magnet of RHEL-Nimrod had to be pulsed and at the commencement of each pulse when 15 KV D.C was applied to cause the current to rise from zero to 9150 Amp in 0.72 s, which was a short rise time compared to other synchrotrons, naturally the peak demand was more. Also, its design of 28 number of pulses pulses per minute was much higher than other synchrotrons that resulted in reducing the total cycle time which was less compared to others. The repetition period could not be made shorter than twice the rise time of the pulse and in case of RHEL, the repetition period was more than twice the pulse rise time. It is also to be noticed that when there was no flat top period for all other units, RHEL-Nimrod introduced a flat top period of 0.115 s which in fact helped to lower the de-energized period. Reason for introducing such flat top in some accelerators has been given in earlier chapter in connection with the findings of Birmingham synchrotron. All these justified the reason for selecting two 60 MW, 11.5 KV alternators for RHEL. Since energy of the rotating masses of the set acted as a buffer, the choice of as high a speed as possible was desired for better performance. Two pole machines with speed of 3000 r.p.m. were preferred because their masses possessed more kinetic energy than machines of equal output with a large number of poles. The selection of salient pole machine had the advantage of its large mass over cylindrical poles, and this need was considered rigorously with other factors in the design. This advantage of higher speed perhaps was the reason for choosing 46 MVA salient pole alternator with 3000 rpm for 25 GeV CERN synchrotron which also utilized the advantage of low sub-transient reactance compared to a multi-pole machine. In choosing the alternator ratings of Cosmotron and Bevatron, the relationship that existed in the case of motors continued which was just double for Bevatron than Cosmotron. As mentioned earlier, the Birmingham synchrotron being the first of its kind in this group preferred to select d.c. generator despite such supply to excite the magnet led to magnet current control problem and invited inherent maintenance drawbacks for its size of 1 GeV.

Flywheels

The rotational kinetic energy largely depended on the selection of the flywheel mass (W) and its diameter (D) which had enormous influence on the capacity of storing energy and supplying it to the generator at the appropriate time. In most cases, this kinetic energy of the flywheel was kept 10 to 20 times higher than the energy stored by the magnetic field when only few percentage of this energy was actually delivered to the generator and converter circuit during magnet excitation. The energy which can be obtained by the rotating masses is proportional to their WD^2 and to the difference between the squares of the maximum and minimum speeds W

(D_{max}^2—D_{min}^2). Therefore, higher the speed difference the smaller will be the flywheel diameter; percentage speed differences were kept nearly 6–7% for alternators with the exception of RHEL where it was only 4.1%. The choice of low flywheel weight of 30 tons for RHEL in this case substantiates the above statement but again, to some extent, the weight of the rotating mass of 60 MW alternators were to be considered as well while comparing with others. The speed difference of such huge rotating sets always confronted with a pulsating load pattern during the short pulse period due to load fluctuation; hence, special attention is given towards damping the vibrations transmitted to alternator. In all synchrotrons, it was necessary to measure bearing vibrations, monitoring of temperatures, shaft eccentricity, torsion stress, and place vibration damping pads on the generator and wherever necessary to limit the transmission of vibration to ensure that any oscillation due to load reversal is sufficiently damped by the time of the next reversal.

Transformers and Converters

Brief explanations on the layout of power plant installation of synchrotrons were presented in the earlier chapter to create some idea on the enormity of the job and emphasize on the elaborate connection of transformer windings and rectifier groupings. The power plant installations with transformer and rectifier arrangements are shown in Chap. 2 for Nimrod and for CERN. In addition to the two synchrotrons mentioned, the layout of the power supply plant of cosmotron and Bevatron is similar to other two arrangements. Standard method of twin supply arrangements of rotating machines with flywheel sets were adopted by some units, but whatever might be the choice of each projects, ultimately, the primary side of the transformers were connected to each alternator and the secondary side connections of the transformers were designed and connected to match the number of rectifier-inverters. It is very clear from the figures that to satisfy the imposed conditions of the supply authority on harmonics, the transformer secondary side configurations were arranged to match with the converter groups connected in series–parallel combinations to supply the demand of d.c. power supply for synchrotron. Actually, all these arrangements were part of the system which converted the alternator output to d.c. for the magnet coil during energizing mode and function as inverters during de-energizing mode. The ignitron type converters for Birmingham after changing from W/L type variable voltage system, Bevatron and Cosmotron synchrotrons were selected for this application, but unusual loading conditions and inverter operations were experienced on such arrangements. Suitability and the adoption of Hg–arc rectifier for Nimrod or for other accelerator unit were under question initially specially due to the problem of backfires. Other units followed transformer and rectifier configurations to suit their requirements to minimize the cost and improve performance.

Ripple Filters

Reduction of ripple in the operation of synchrotron operation is a wide subject and it will be interesting to know where from the ripples could be generated which ultimately could affect beam stability. The following specific areas are mainly identified as the origin of ripple generations and corrective measures are taken to minimize ripple in selected zones:

(i) voltage ripples in power converters,
(ii) current ripples due to magnet inductance,
(iii) magnetic field ripples affecting flux densities.
In RHEL, the ripple frequencies were higher than the frequency components of the basic pulse for which the installation of filter network was quite practical. Necessary efforts were made to minimize ripple content in the rectified d.c. to keep it within limits so that operation of the magnet was no way affected. Similar attempts were made in rectifier circuits of all other accelerators to keep the ripple content minimum to prevent the proton beam to modulate and give rise to undesirable effects. One method of reducing this ripple effect was to interpose a suitable filter between the rectifier and the magnet. In their design of low level power supplies for electronic circuits, a large capacitor at the output of the rectifiers is common to reduce the ripple content. Conventional filters between converters and magnet were straightforward in principle but large in size because the series elements had to carry heavy currents. The primary system employed on 24-phase rectifiers in Nimrod had inherent ripple component superimposed on their output voltage, and this ripple power was high. It consisted of two amplifiers which produced a voltage in phase opposition to the ripple voltages. The smoothing circuit of a simple LC element was used in CERN. Hg–arc rectifiers and ignitrons were declared obsolete nearly 50 years back due to emerging technology of semiconductors when thyristors were accepted as the most suitable replacement as a converter device but the fact is that the transformer and rectifier configurations and ripple reducer arrangements remained same for which due weightage is given in the design to reduce the ripple content. But again, demanding high performance through ripple reduction is tied up with additional cost, but this may become a potential source of component failures and a burden for maintenance which limits the design.

Electrical Distribution System

Distribution and utilization of electrical power at the Rutherford high energy laboratory offered a comprehensive view of the power supply arrangements provided to a High Energy Physics Research Station with synchrotron installations (Dey,

1969). The possibilities of considering similar scope of electrical system design but of different ratings for the electrical distribution for other projects may also help in the design (Gashev, 1956). The most important part in the design of power supply was to ensure the continuity; this was the reason for which the supply was taken from separate sources. The variation in the design was due to different synchrotron parameters and the services required for other experimental setups which dominated the selection of the type of power distribution network. Enormous amount of power was required for focusing, bending, and for steering the beam in Nimrod. In RHEL, the total power requirement for the beam line magnets was of the order of nearly 10 MW and the cost of such high power requirement was considered equivalent to the cost for beam transportation. The cost of such beam transportation was more or less same for the other synchrotron units depending on energies and peripheral equipment. In order to protect the equipment from the unwarranted influence of supply disturbances, a strong network design was implemented in all accelerator facilities. The maximum demand of electricity at the RHEL was in the region of 18 MW of which as a rough estimate 7.5–10 MW was required to pulse the Nimrod magnet, 1 MW to PLA, 4 MW for the bubble chambers, and the remainder was required for other auxiliaries such as r.f supply, air conditioning, cooling system, vacuum, lighting, workshop facilities, and high voltage testing labs. The basic considerations in the design of distribution system were to

(a) supply electricity to the experimental equipment for the production of proton beam,
(b) supply power to the experimental areas and auxiliary supply for machine operations,
(c) supply lighting system and other working areas.

Workshop Requirements

It is worth mentioning that, most of the complex scientific tools required for experimental purposes are manufactured in the laboratory's workshops. This has several advantages. Firstly, continuous inspection, guidance, and monitoring to satisfy the prescribed precision is available at minimum cost. Secondly, the waiting period of contracting and delivery is avoided. Thirdly, it leaves a scope for modification and any design change at a later stage. At Rutherford Laboratory, engineering group had its main central workshop supported by medium-size electrical/mechanical workshop at each department. Most accelerator facilities followed the same and developed their own engineering workshops to carry out urgent experimental requirements. The load requirement of such a workshop was of the order of 120 KW.

Lighting Load

The total lighting load of the laboratory was approximately 0.5 MW which was considered high. The main contribution to such high demand of lighting load was due to high level of illumination in the magnet hall, bubble chamber area, and the experimental zones.

The most popular method of load equalization is achieved by employing duplicate bus bars. In this arrangement, secondary winding from each transformer is associated with each bus bar and load equalization is achieved by transferring feeders from the front to the rear bus bars or vice versa as required. The weakness of the arrangement using duplicate bus bar is the connection of switchgear inherently off load. The primary purpose of bus bar load equalization was to achieve equal voltages on the 11 KV bus bars. Automatic tap changers on the transformer windings were mainly to keep the bus bar voltage within the limits at all times. It is obvious that a design of sub-station switchgear enabling immediate or on load transfer of feeders would be of advantage.

Summing Up

It has been more than 70 years the synchrotrons were developed and the research continued throughout the world vigorously to achieve the goal set to advance the scientific studies with experiments. All those old synchrotrons were fully dismantled to give space to new methods; modifications are carried out for further research to explore new possibilities. New methods are being tried to reduce the dimension of the huge magnet and sort out the stored energy problem to minimize the investment and operational cost. Achieving high reliability, increase lifetime, expand mean time between failures of critical components, selection of suitable equipment to control over sizing of power supplies, provision of redundancy measures are major technical aspects that need to be seriously evaluated and compared against the cost of breakdown due to faults or major stoppages of the accelerator causing setbacks for the experimenters. The concept of continuous power supply and robust grid for direct power supply is also in the list. Research on superconducting magnets are continued vigorously to reduce the size of the magnet and for other applications. High energy physics is notable for the application of superconducting magnets for economic reasons because the high magnetic field would allow large reduction in the circumference of the magnet ring. One other interest is the possibility of superconducting magnets for bubble chamber which consumed large amount of magnet power.

References

Dey PK (1969) Distribution and utilization of electrical power at the Rutherford high energy laboratory. SQJ IEE (Lond):1–7

MA Gashev (1956). Power supply system of the 10 GeV synchrotron electromagnet. Proc. CERN, Symposium of high energy acceleration, p. 378.

Green GK (1953) Cosmotron magnet power supply. Rev. Sci Instruments 24:769

Kresser JV (1952) The Bevatron Power Plant. Electr Eng:338

Chapter 4
Changing Patterns of Magnet Excitations and Energy Storage Schemes for Particle Accelerators

Bottlenecks of Earlier Synchrotrons

At the initial stage of development in 1950s, the most dependable technology for the magnet excitation of the operating synchrotron accelerator used flywheel as energy storage device and how effectively this scheme was implemented and utilized has been discussed earlier. In the absence of any alternative for powering magnet excitation, the above scheme was considered as the final and the best design choice at that time. But, the structure of the selected system unfolded few serious bottlenecks during the course of operation of accelerator which were not taken into design consideration earlier due to lack of proper technical inputs and other reasons. The complex scheme of earlier design shown in Fig. 4.1, essentially had two parts: rotating parts—comprising motor–alternator–flywheel and stationary parts comprising—transformers, rectifiers, and magnet systems arranged in tandem for supplying pulsed power to the magnet coil. It was a challenging task to the scientists and technologists during the middle of the twentieth century to ease out from the inherent bottlenecks of mechanical energy storage and pulse power supply and find an alternative method of supplying power for magnet excitation to comply with the various demands from projects experiencing operational difficulties. Magnet power supplies for Nimrod or similar accelerators were designed to operate for long periods with a very high reliability. The plants were designed with minimum life expectancy of 15 years, but due to repetitive exchange of energy between the generators and the magnet coils, the equipment were subjected to high mechanical, thermal, and electrical stresses which ultimately affected the performances. The urgent need of a new concept was purely to avoid the bottlenecks for the reason that earlier system had to work under various restrictions and imposed conditions, some of which are as mentioned below:

P. K. Dey, *Power Supplies for Synchrotron Accelerators*, Springer Tracts in Electrical and Electronics Engineering,
https://doi.org/10.1007/978-981-97-7606-1_4

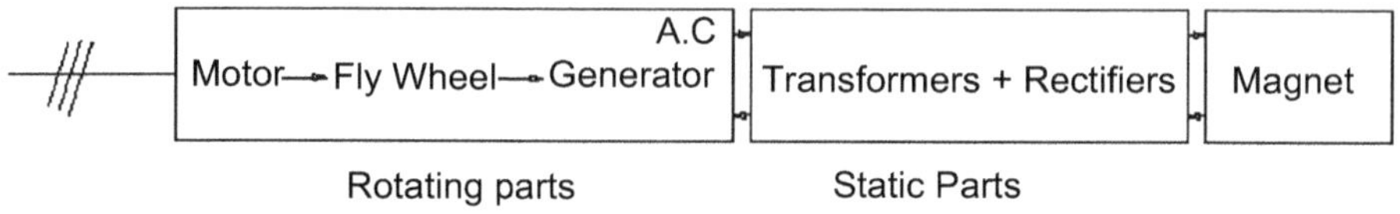

Fig. 4.1 Schematic of magnet system

(a) Uniform loading of the supply network without surges.
(b) Prevention of active and reactive load surges from passing to other networks or consumers.
(c) Magnet voltage to remain within the limits of ±2% during magnet excitation.
(d) Change of alternator speed during initial to peak period of magnet excitation within set limits.
(e) Strict monitoring and control of shaft vibration and eccentricity measurements of alternators to minimize breakdowns.
(f) Suitable alternatives for Hg–Arc rectifiers to limit operational losses and maintenance problems.
(g) Strict monitoring and control of r.f. generation with magnetic field variation.
(h) Maintaining correct vacuum inside the torus of the magnet.
(i) Nonavailability of critical spares in time put stress on cost to make operation prohibitive.
(j) Apart from that, high power loss, high reactive power consumption, low power factor, and severe vibrations transferred to rotating equipment due to pulse loading were some serious drawbacks to the mechanical energy storage system needs to be improved.

In search of viable alternatives, extensive analysis on various schemes was carried out to find a feasible and acceptable solution for supplying power for magnet excitation. During evaluation process to find a new technology for supplying excitation current to magnet, the subject of raising beam intensity was also given some special thrust and considered as priority for which, the following criteria were also included in the search.

(i) Inject large current into the torus of the magnet by increasing the initial energy level.
(ii) Increase the repetition rate of operation.
(iii) Inject the particles for as long as possible.

Origination of New Concept

The complexity in the design and operation of power supply for magnet excitation played a significant role in lowering the performance of the earlier synchrotrons which consumed nearly 70 % of the total energy. To give some idea, the electricity bill of the Rutherford Laboratory in 1965–66 was nearly as high as £ 500000 a year. Other operating synchrotrons also faced more or less similar exorbitant energy bills at that time. Apart from various reasons mentioned earlier as bottlenecks, excessive energy bills topped the most and caused serious concerns for lowering down the performance and made economical operation of accelerator difficult. Some other factors under scrutiny were the technology adopted for magnet excitation, the choice of pulse repetition rate and increasing demand of time average beam intensity. It became important to explore on the development of an alternative method for a change in the pattern of magnet excitation on priority for different categories of accelerators to generate high intensity beams with the inclusion of criteria mentioned earlier and improve performance level with minimum of electrical consumptions. A befitting solution to this challenge was ultimately offered by changing the functioning of the magnet coil orientation in the pattern of magnet excitation. Prof. M.G. White conceptualize a change in viewing the magnet circuit differently by considering the magnet coil simply to be divided into small sections of equal impedances connected in series and transferring inductive energy stored in each at its peak to a capacitor bank in resonant magnet configuration mode. Enormous possibilities evolved from this simple concept which was considered to be a great technical achievement. This idea was soon accepted and utilized as a case study in developing 3 Bev Princeton-Pennsylvania synchrotron accelerator (White et al. 1956) which ultimately became pioneer and the benchmark model for the development of subsequent accelerators. At that time, successful implementation of such pioneering concept in a synchrotron project was a relief from various operational difficulties and considered to be a quantum jump from mechanical energy storage to inductive-capacitive energy storage, from pulsed power to cycling power accelerators with substantial reduction in electrical consumption. The Inductive energy storage ($1958–1998^{+}$) adopting principle of "White Circuit," proved to have various advantages over mechanical energy storage and set a new direction for resonant circuit magnet excitation which slowly gained momentum in its choice for operation and ultimately made the earlier system of mechanical energy storage scheme and supplying exciting power to magnet obsolete by 1970s. This new found alternative technology was also adopted for 4 GeV electron synchrotron of Daresbury Laboratory (UK) set up by the NIRNS (National Institute of Research in Nuclear Science UK) as their second unit (Fox 1965) (the first unit being the NIMROD of Rutherford High Energy Laboratory, RHEL) to reduce the reactive power consumption and improve power factor during operation. The dynamics of resonant network and the economics involved in the selection of components opened up new application possibilities in magnet excitation to strengthen Prof. White's achievement and ascertained subsequent advantages offered by this technology. The "Single cell" configuration in its simplest form shown in Fig. 4.2, named after its inventor Prof. M.G. White, has changed the magnet orientation and divided it into small sectors where each magnet cell forms resonant circuit of its own. The number of such single cells connected to form a series string of the accelerator magnet with its collective inductance L_m and lumped capacitor C_m in parallel forms a resonant network. Another circuit, supplying a.c sinusoidal waveform, at required operating frequency,

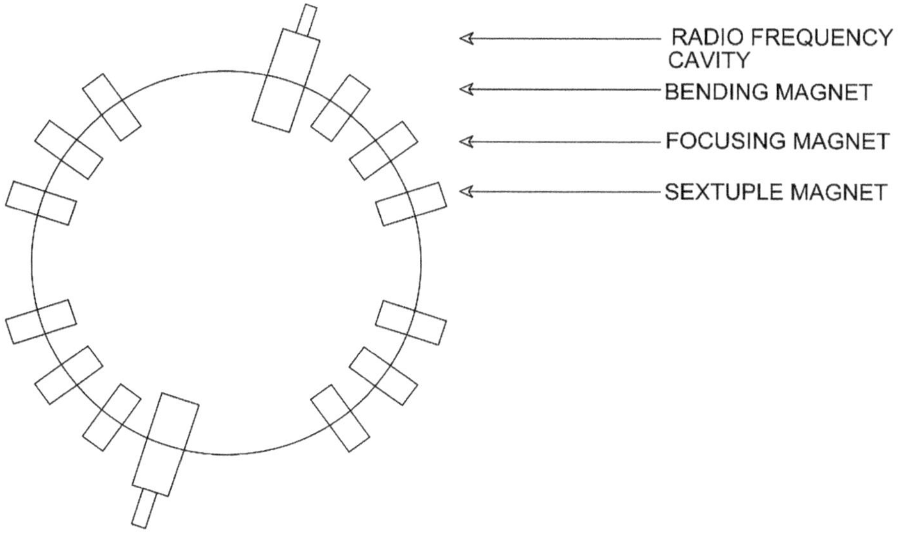

Fig. 4.2 Magnet orientation in "Single cell" White circuit

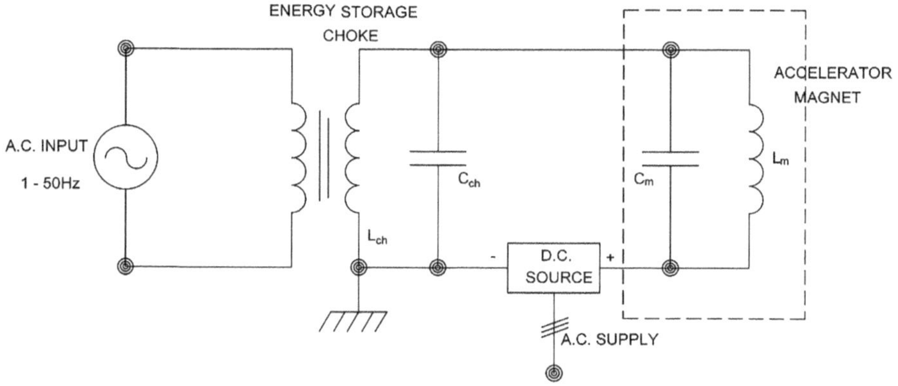

Fig. 4.3 An electrical schematic diagram of White circuit

vital for White circuit is fed to the primary, coupled to the secondary, referred as the inductive energy storage choke L_{ch}, in parallel with the capacitor C_{ch}, forms a separate resonant circuit. The two isolated resonant circuits are connected by a d.c source to satisfy the demand of high acceleration voltage to inject large current into the magnet by increasing the initial energy level required at the point of injection for normal sine wave excitation as shown in Fig. 4.3. Two critical magnet field parameters which occur once per cycle during the particle acceleration period are as follows:

(i) The low positive field gradient at which particle injection takes place at the commencement of the cycle specifically, at the time of injection is the field B_{inj}, which is maintained within ±1%. As mentioned earlier, the initial injection energy corresponding to initial field value has to be of some critical value, higher than the threshold for the particle to be accelerated inside the torus of the magnet.
(ii) The other one is the peak field amplitude B_{peak}, which is maintained within ±0.01% at the end of acceleration period when the particles attain the maximum energy.
During these periods, high stability and extreme precision of the magnetic fields are considered to be the key factors for accelerator operation; therefore, the value of B_{inj} and B_{peak} are to be properly selected and their accuracies are to be maintained within limits when developing or upgrading the design of magnet.
As it can be seen in the figure, the dominant features are two types of power supplies, a.c and d.c required in the operation of "White circuit."
When a.c supply to the circuit is chosen from three basic types as follows:

(a) Continuous excitation from a rotating machine, in which a single frequency controlled alternator, is connected to the choke primary windings
(b) Continuous excitation from a resonant circuit converter
(c) Intermittent excitation from a pulse power supply

On the other side, the d.c power supply for the field biasing was chosen for the following reasons:

(i) The introduction of a d.c. source in the circuit is to provide a path for the d.c. bias current to minimize the circulating alternating current.
(ii) D.C field bias approximately equal to a.c amplitude is suitable for the resonant magnet circuits which materially reduces the required capacitor bank and lead to substantial savings in power dissipation.
(iii) The biased sine wave provides a lower maximum value of $\frac{1}{B}\left(\frac{dB}{dt}\right)$ at injection compared to linear ramp (Bouteille 2015).

The combined resonant frequency of the parallel resonant circuits shown in the figure is represented by

$$\mathbf{f_r} = \frac{1}{2\pi}\sqrt{}\left(\mathbf{L_{Ch}} + \mathbf{L_M}\right) / \mathbf{L_{ch}}\mathbf{L_M}\left(\mathbf{L_{ch}} + \mathbf{L_M}\right) \tag{4.1}$$

It is clear from the above expression how important is the selection of inductors and capacitors in the design, as the drifting values of components will affect the resonant frequency which ultimately results in disturbing magnet excitation. In resonant condition, the beam intensity is raised to maximum during acceleration and the stored energy in magnet is transferred to the choke during deceleration. As mentioned d.c and a.c cyclic waveform are two separate stand-alone supplies and essentially required for resonant magnet excitation to express the accelerator magnet current in the form

$$\mathbf{I}_{\mathbf{mag}} = \mathbf{I}_{\mathbf{d.c}} + \mathbf{I}_{\mathbf{a.c}} \cos\left(2\pi\, \mathbf{f}_{\mathbf{rep}} \mathbf{t}\right) \tag{4.2}$$

During resonant condition, the total network stored energy is constant, excepting losses where, the a.c. power makes up for the losses for the system and the magnet stored energy oscillates between the lattice magnets and the energy storage chokes during the acceleration and deceleration periods once per cycle with the capacitors storing a portion of peak energy during the intermediate stage.

This is possible when the current is initially injected to the magnet at low field and subsequently raised till the field reaches its peak value and the energy so stored in the magnet is removed during the deceleration phase. The essential parameters of magnet current requirements are its stability, reproducibility, and accuracy for improving the operational efficiency and susceptibility which is different for different machines. The flexibility of waveform adjustments are preferred to match the beam requirement during acceleration but, to avoid drift in the quality of beam, special emphasis is given on the design, monitoring, and control of parameters. The a.c source is either fed via coupled auxiliary windings on the energy storage choke (parallel resonance) or via series connected power transformers in the resonant-capacitor groups (series resonance) to meet the active power losses of the magnet circuit. In fast cycling synchrotron, three components: power sources, magnets, and storage energy devices are important in designing a multi-cell resonant circuit configuration. In such applications, both parallel and series resonant circuits are suitable and selected after evaluating the following differences:

- In PRC, two separate sources are required; the a.c current source and d.c voltage source are separate and the current is fed in the choke and not directly fed to the magnet. Peak a.c. voltage is supplied and operated at low power. The fault protection of the parallel connected power sources is much simpler.
- In SRC, only one serial connected current source provide a.c and d.c and the current is directly connected to the magnet. Peak a.c voltage and the peak power drawn is more in SRC during a cycle.

The scientific research continued in new directions to increase the number of charge particles per pulse, upgrade for more charge and high intensity beams. But, requirements of high beam energy imposed stringent demands on stability, efficiency, accuracy, and electromagnetic compatibility of magnet system during operation which invited again a separate design challenge. Economic evaluations are required to emphasize how simple, reliable, and cost-effective is the changed process.

Some classification of synchrotron accelerators have been presented earlier but, to identify and make the search simple of a specific type, particle flow sheet of a typical synchrotron accelerator system is shown in Fig. 4.4. However, it is important to mention that since classification of particle accelerators based on physical

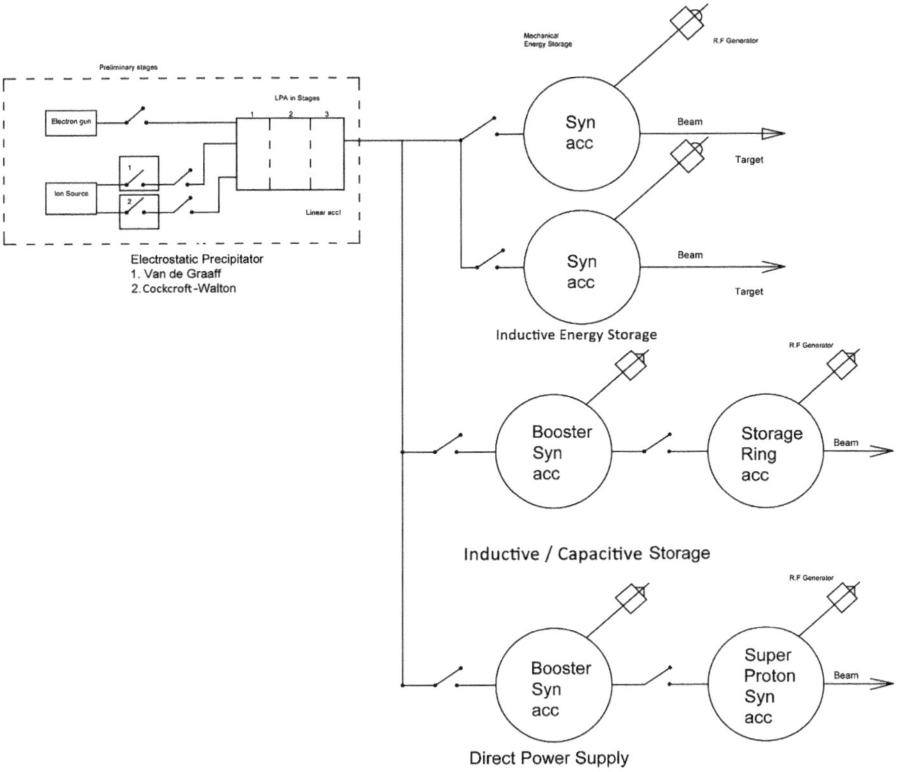

Fig. 4.4 Charge flow sheet of synchrotron accelerators

principles has been dealt elsewhere in various literatures, this part is not considered here. The model presented here has three preliminary stages including two streams of handling ion sources and linear accelerator stage mentioned earlier being part of the system is marked separately with dotted lines. Apart from that, the figure indicates the pattern of magnet excitation and clubbing of synchrotron accelerators for different machines in terms of energy storage, cycle frequencies with following boundaries:

(i) In the operation of a large synchrotron, parameters like frequency of operation and the type of the waveform play an important role.
(ii) Synchrotrons with normal linear mode ramping for excitation are not suitable for higher repetition frequencies.
Excitation of synchrotrons having higher repetition frequencies ω runs into difficulties if a normal linear mode ramping is used because the electrical power for energizing and de-energizing the magnet increases with ω. Working limitations on the operating synchrotrons and the demand for higher intensity beam evolved this concept of new magnet excitation where the magnetic fields are required to be raised in a relatively short time from the low amplitude at beam

injection to maximum corresponding to peak beam energy and thereafter, fields are withdrawn in as short a time as possible, so as to be ready to accept the next pulse of injected beam. Actually, to achieve high pulse rate, low peak magnetic energy needs to be stored only for a short time by reducing the accelerating period, and this is possible by changing the concept from pulsed power to cycling power with higher frequencies for magnet excitation.

Power supplies are designed according to their frequency of operation and following terminologies are used to identify the specific accelerator as shown in Table 4.1.

For cyclic application, alternating current waveform is preferred but again, as mentioned sine wave without a d.c bias is not desirable because, it is effective only during positive cycle when the magnetic field variation is high between injection and peak energy. In comparison to simple a.c for magnet excitation, with the inclusion of d.c bias, the increased duty cycle reduces the field gradient and magnet core losses. Thus, a.c with a d.c bias of amplitude $I_{a.c}$ and $I_{d.c}$, both independently controlled is found suitable for the application in White circuit. The field variation during excitation of the magnet with a.c sinusoidal waveform superimposed on a d.c bias is

$$\mathbf{B_f(t) = B_{d.c} - B_{a.c}\sin(\omega_a t)} \tag{4.3}$$

And the field ramping with d.c biased a.c power supply is expressed as:

$$\mathbf{B = 2\pi f_{rep} B_{inj} \sin\left(2\pi f_{rep} t + \pi/4\right)} \tag{4.4}$$

Where f_{rep} and B_{inj} are the pulse repetition rate and field amplitude at injection, respectively.

The acceleration rate follows the ramping rate of the sinusoidal magnetic field and the maximum acceleration occurs at the end of a quarter of the whole cycling period, which is significantly larger than a linear field ramping. On the basis of biased excitation, the value of magnetic current with time is given by:

$$\mathbf{I_m = I_{d.c} - I_{a.c}\sin(\omega_a t)} \tag{4.5}$$

Table 4.1 Accelerator terminologies

Speed	Cycle frequency (Hz)	Time	Synchrotron project
Slow cycle accelerator	0.1–1.0	8.94 s	300 GeV CERN Super Proton synchrotron
Medium cycle accelerator	1–10	100 ms	3.0 GeV European synchrotron radiation
Fast cycle accelerator	10–50	20 ms	4.0 GeV NINA, Daresbury accelerator (UK)

The power flow in cyclic system depends on the requirement of beam energy which is proportional to the magnetic field generated by the current circulating in the magnet coils:

$$\textbf{Stored Magnet Energy}; \mathbf{E_m} = \frac{1}{2}\mathbf{L_m}\left(\mathbf{I_m}\right)^2, \tag{4.6}$$

$$\textbf{Power flow}\left(\mathbf{P}\right) = \mathbf{V_m}\mathbf{I_m} = \frac{\mathbf{dEm}}{dt} + \mathbf{R_m}\left(\mathbf{I_m}\right)^2 \tag{4.7}$$

P = (Rate of change of magnet energy) + (Resistive power loss)

Where L_m = inductance of the magnet coil

R_m = resistance of the magnet coil

V_m = voltage of the magnet coil

I_m = magnet current

Some typical magnet parameters of particle accelerators with cycle frequency (speed) are shown in Table 4.2.

With cyclic power, the main challenge is to provide and control the positive and negative flow of energy as field rises and falls on their frequency of operation. Increased beam intensity demands high rate of change of field. But, fast pulsing imposes severe tolerances on injection, r.f.g (radio-frequency generation), magnet design, vacuum, and heat removal systems.

Radio-Frequency Generation and Control System

It has been mentioned earlier that magnet system and radio-frequency generation and control are complex part of synchrotron operations. So far, only a brief description of r.f supply with reference to Nimrod as an example has been given in Chap. 2, but this subject needs to be discussed further to stress on some critical points in the complex scheme of design of such r.f.structure. Actually, the function of r.f g is to impart energy to the particle in r.f cavity placed across gaps at different points as per the need of the accelerator to boost up particle energies during each revolution. Two types of r.f.g is available for selection, one is ferromagnetic core which is excited by an alternating current whose frequency during the acceleration period

Table 4.2 Typical magnet parameters with frequency

Speed-cycle frequency	Magnet resistance	Magnet inductance	Magnet stored energy (MJ)
Slow cycle (SPS) 8.94 s	3.25 Ω	6.6 H	109
Medium cycle (ESRF) 100 ms	565 mΩ	166 mH	209
Fast cycle (NINA) 20 ms	900 mΩ	654 mH	606

varies and the other one is tuned system, i.e., a resonant cavity (Klystron) driven at varying frequency and continuously tuned over the desired frequency, which is the result of the new development in the design of r.f supply. The new concept of resonant cavity is the foundation of microwave technology developed to improve the cavity design after the experience gained from World War II to increase the operational frequency and r.f power. The basic function of r.f.g is the same as explained for Nimrod, and for all other accelerators the method adopted may be different where modification is introduced with advanced system. After the injection of the particles into the torus of the Nimrod synchrotron, the particles followed a circular path guided by the magnetic field. While following the circular orbit, the particles pass through accelerating field which raises the energy level gradually during every rotation from the entry level energy to the maximum value. This process of gradual energy transfer in every cycle took place in specially designed r.f. system for Nimrod, where the electrical energy from the grid controlled triodes was transferred to the r.f. cavity which in turn raised the energy of the accelerated beam. It has been pointed out earlier that the most challenging task is to synchronize the magnetic field with the variable accelerating frequency. It was indeed difficult to vary the magnetic field and the accelerating field simultaneously to realize an exactly predetermined increase of the magnetic field with time due to the magnet being a nonlinear element in the circuit. Figure 4.5 also shows the variation of flux density and frequency with time during one acceleration period. The particles move in their circular orbit due to the magnetic field and considered as independent variable, while the frequency of the accelerating electric field vary according to the curve f, at an increasing rate to ensure that the circulating particles remain inside the torus

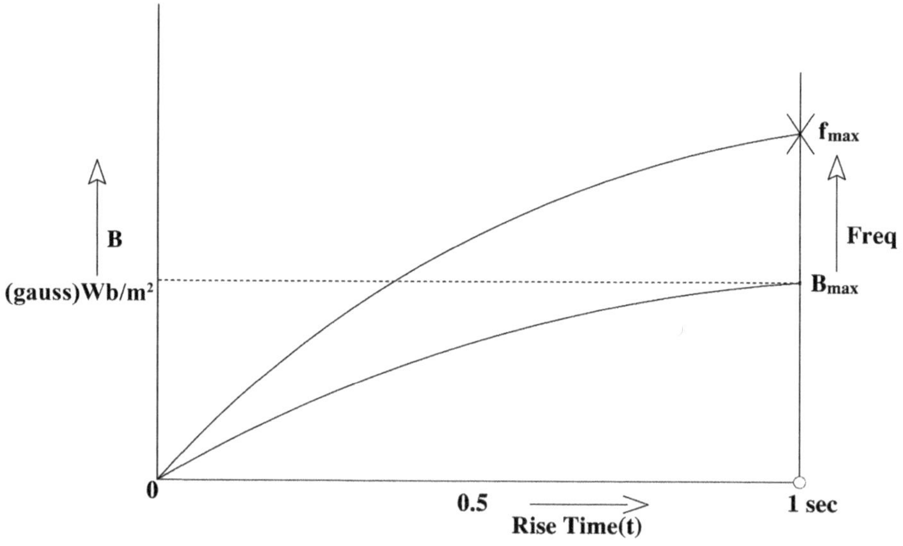

Fig. 4.5 Variation of B and f with time. drawing fig 3 0

of the machine. An expanded schematic diagram of a typical r.f. system shown in Fig. 4.6 emphasized the essential elements required for developing a suitable r.f. system. Although powering and controlling for r.f circuit comprising the two most important sections could be seen as separated, both were equally important in the formation of the basic r.f structure. Arrangements of the expanded circuit diagram of r.f.g has been described in simplified form and shown in Fig. 4.7. The operation of the r.f.g in synchrotron depended on their contribution in controlling and maintaining the accuracy and stability of the particles where

(i) The first part consisted of one high power 3-Ph high voltage rectifier unit for converting a.c to d.c which supplied power to r.f. cavity responsible for providing the accelerating voltage.

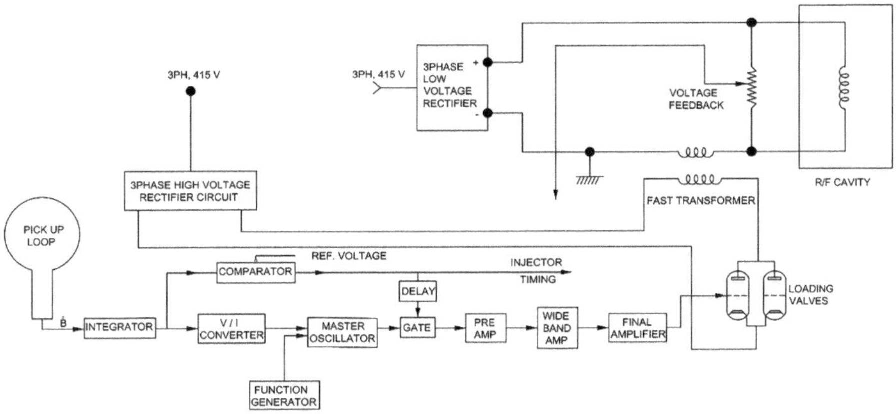

Fig. 4.6 An expanded schematic diagram of a typical R.F. system

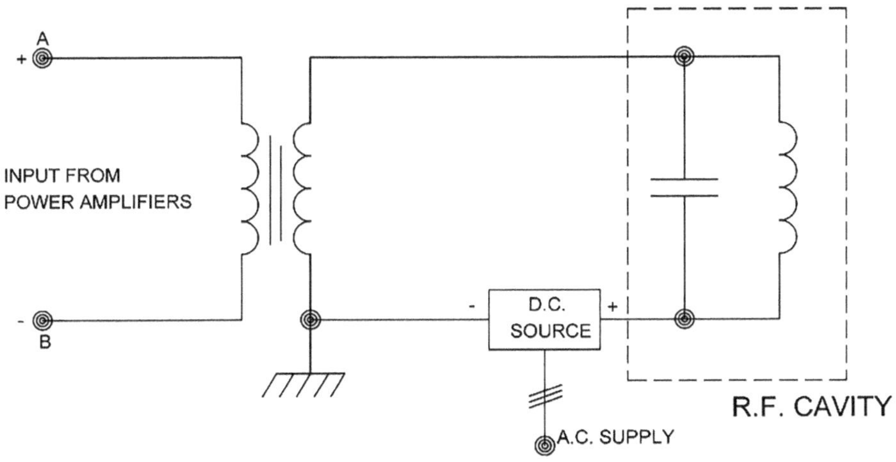

Fig. 4.7 A simplified form of R.F. circuit

(ii) The second part consisted of a low power 3-Ph voltage rectifier unit to supply power to the low level r.f unit and provided an accurate and controllable frequency source for the high power r.f system. The frequency had to be generated accurately at low power and then further amplified with additional circuits. Low level r.f unit in a sense functioned as the brain of an r.f system because it measured the primary information from the magnet and controlled the flow of information required for providing a stable operation. Also, it helped to interact with operators who set the system parameters and controlled it as per the experimental program. A typical timing sequence for one complete pulse period on a logarithmic scale used in RHEL, Nimrod is shown in Fig. 4.8. This timing sequence was strictly followed by the operators of RHEL where other synchrotron units also developed similar time sequence of their own to satisfy the requirement for smooth operation.

In order to solve the problem of adjusting the magnetic field with the accelerating field simultaneously, the control circuit was designed in such a way that the derivative of the magnetic field signal had to be monitored continuously by means of a pickup coil and further processed through amplifiers in the electronic circuit to control the operation of r.f generation. The dB/dt signal received from the search coil was passed through an integrator, a buffer amplifier, master oscillator, and a set of r.f amplifiers. To keep the voltage and frequency of r.f system under control, a number of feedback loops were required in earlier synchrotrons. The function of the amplifier chain was to amplify the signal to higher power for which this low level signal was connected to the grids of a pair of triodes supplying r.f power. One of the important tasks of r.f amplifier was to convert d.c to a.c power in an active element to control the frequency of an LC oscillator producing the accelerating frequency.

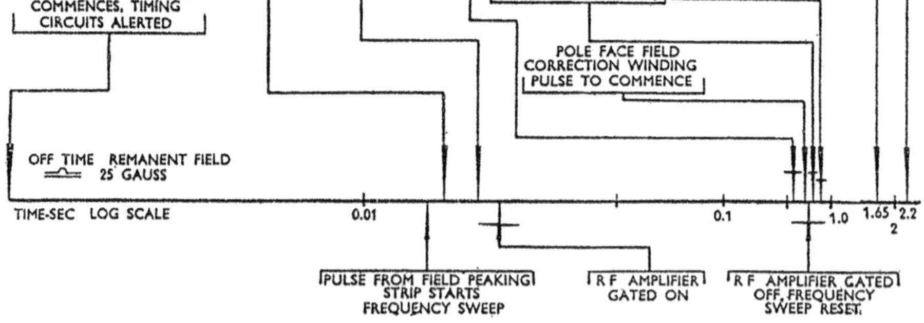

Fig. 4.8 A typical pulse sequence time for Nimrod

Such conversion took place when the d.c voltage was modulated at r.f frequency and the r.f cavity a resonant structure formed the load across the power triodes. In fact, the r.f system transferred electrical energy received from the grid to the beam of particles. Thus, r.f cavity in which an axial electric field is established alternates in synchronism with the proton orbital frequency and this electric field is followed and maintained:

(i) the correct frequency with respect to the particles
(ii) kept the correct phase, with respect to the particles
(iii) always maintained the correct amplitude
Actually, the origination of induction synchrotron concept during the operation of r.f synchrotron was for few limitations such as:

- Band width limitation – This is because resonant frequency of a resonant cavity and r.f amplifier is expected to change slightly.
- Limited acceleration phase – The acceleration phase is a region of positive voltage growing in value over time.
- Localized higher line density – An essential aspect of the phase stability tends to increase the local beam intensity around the bunch center.

$$\mathbf{R.f\ acceleration\ voltage;\ V_{r.f} \propto dB / dt} \quad (4.8)$$

Since the acceleration voltage is related to magnetic flux density, dB/dt limits the maximum value of radio frequency. It has been mentioned earlier that power to produce the required radio-frequency electric field on particle beam inside the cavity is derived from high power amplifiers shown as part of expanded circuit with control system of a typical r.f energy transformation process. The heat transfer technology in the r.f system demanded most sophistication, and this complexity existed in the design, fabrication, and also in operation. This is best explained when the functional structure of the r.f system is viewed as a black box where the energy from the external source raised the low input energy of the particle beam to acquire higher output energy in every cycle as explained and shown in Fig. 4.9. When low beam energy "W" is passed through the accelerating cavity, it gains energy quantum "dw" and thus become "W+dw" at the exit and this increment continues in every cycle.

$$\mathbf{R.f\ power = (Power\ transferred\ to\ beam) + (Energy\ Loss\ into\ cavities)} \quad (4.9)$$

The r.f. energy imparted on the beam undergoes three-stage transformations which lowers the efficiency of r.f. system and when the number of r.f stations is more, cumulative losses increase. Energy losses in power triodes and the cavity are extracted by suitable heat transfer arrangement with the objective to reduce high power losses of the r.f system and improve the performance. Efficient heat removal and utilization of extracted heat from the system is therefore essential part of the design.

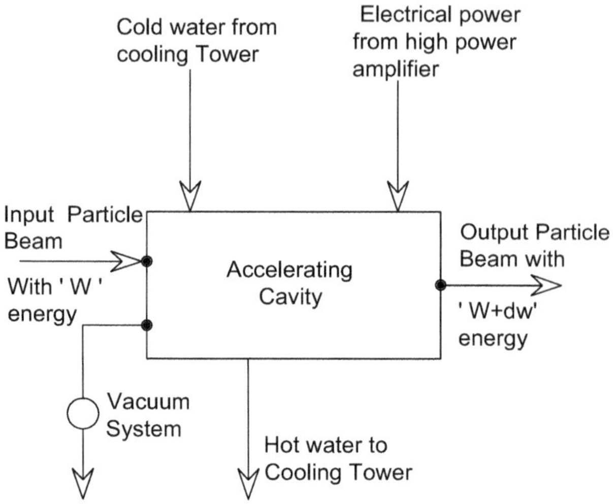

Fig. 4.9 Black-box representation of R.F. circuit

Similarities and Differences—Comparison of Schematics; r.f. Generator and Resonant Magnet White Circuit

It may raise question marks to know what advantages the new scheme could offer and what might have led to the origination of this new concept and how this ingeneous approach of attempting to divide the magnet coil in small sections and form single cell structure to utilize the principle of inductive energy storage came to the mind of Prof. White at that critical time when all the scientists and engineers were anxiously searching for an alternative. It may be interesting to make a wild guess, a pure assumption that this concept probably materialized when Prof. White was concerned with the exorbitant electricity bill for operating accelerators and placed special attention to find how to clamp down large reactive power consumption which contributed to maximum extent in lowering the performance of the earlier synchrotrons. Due importance was given on the number of resonant cell structures at the time of designing a resonant circuit based on the new concept for magnet excitation supply particularly and considered the following factors:

(i) The importance of maintaining equal current and uniform field
(ii) The selection of operating frequency
(iii) The selection of energy storage chokes
(iv) The selection of energy storage capacitors
(v) Type of d.c converter supply and control circuits. The idea of field biasing the magnet to reduce the requirement of capacitor of suitable rating, which will lead to substantial savings in power dissipation
(vi) a.c supply configuration and applied waveform for magnet excitation
(vii) Direct connection source for magnet excitation to network with high Q factor which exhibits purely resistive load to the a.c power supply under symmetrical resonance condition

Today, the perception of Prof. White may sound simple and easy to suggest such a scheme but, in late 1950s, this seemed to be the toughest hurdle to overcome the problems which seriously affected the performances of accelerators. In this context, with full respect to acknowledge Prof. White's achievement for his great perception in dividing the magnet coil to the number of cells to hold lower energies per unit, selecting cycling power in place of pulsed power, resonant configuration and d.c biasing, which brought a revolutionary change in the structure and operation of accelerators and the total approach to this possibility need some searching for the origination of this superb idea. Most interesting part is the introduction of d.c bias in the magnet excitation circuit which has proved later to contribute enormous advantages over the conventional circuit. The quantum of improved efficiency has been shown later with examples of the operating accelerators. A wild assumption after 50 years of the origination of this idea of magnet excitation which resulted simultaneous reduction in the consumption of reactive power and improve power factor would possibly be that Prof. White might have been inspired to develop the new concept from the circuit design and functioning of r.f. generation. Such hypothetical assumption originated in the mind of the author may be totally wrong and the readers may do some research and look into some other possibilities but, some facts lead to suggest that, r.f.g circuit might have aided to introduce a new perception for the following reasons:

(a) Similarities in schematic diagrams

It is interesting to compare the schematics and brief circuit diagrams presented for r.f.g in Fig. 4.7 and White circuit in Fig. 4.3 which exhibit apparent circuit similarities between the two when placed side by side and viewed suddenly with the absence of input parameters. Actually, the circuits reveal their functions to be widely different when the operating frequencies and inputs are introduced in the circuit diagrams. The interesting comparable features of the two circuits to note are that

(i) r.f.g is a simple ordinary circuit with resonant structure and White circuit is also a resonant magnet circuit.
(ii) The first one develops and controls electric field because the basic function of the r.f system is to provide the correct electric field, whereas the second one develops and controls magnetic field. The dominant feature of the "White Circuit" (resonant circuit) is to develop correct magnetic field during the point of beam injection to the point of beam extraction, respectively.
(iii) The input frequency to r.f.g is high while the input to "White circuit" operates between 1 and 50 Hz depending on the type of accelerator.
(iv) In both the circuits (r.f.g and White), a d.c bias is required. The high power part of the r.f system comprises the accelerating cavity which forms the load across the power triodes and is a resonant structure.
(v) In r.f.g, the output of the thermionic valves controlled by their grids transfer high r.f power to the coupled inductor which is connected in series

with a d.c. source feeding the ferrite loaded accelerating cavity. The bias current varies the permeability of the ferrite and thus tuning of cavity is accomplished by a biased magnetic field. The r.f electric field during an acceleration period alternates in synchronism with the protons orbital frequency to maintain correct frequency, phase, and amplitude which is a function of the rate of momentum of the particles. In "White circuit," the application and the advantages of d.c biasing in a resonant network has been discussed earlier in the text. The power converters needed for White circuit has to be of high precision and maintain high accuracy for the magnet current. In this case, the central timing sequence developed for the particular accelerator controls all the magnet power converters and other supporting devices required for acceleration.

(vi) r.f.g requires close control of amplifiers after receiving the feedback signal of magnet flux density from high quality pick up coil and monitoring of the timing sequence for one complete pulse period to maintain the accuracy and stability of the particles. In "White circuit," to achieve high pulse rate, accurate feedback signal of magnet current is essential to maintain high stability, tracking, monitoring, and control of resonant circuits.

(vii) In case of r.f g, the energy input in eqn. (4.9) is in two parts, the first part kicks up the acceleration voltage by transferring energy to the beam and the other part is the energy loss in the process. In case of "White circuit," as in eqn. (4.7), the first part is the rate of change of magnet energy and the other part is the energy loss in the magnet.

(viii) In r.f.g, higher frequencies transferred from the power amplifiers are used in the resonant cavity, whereas in "White circuit" it is a resonant magnet circuit and the essence of the design is to treat the magnet coil in small sections for effecting better handling of the circuit thereby lowering the power consumption.

So far, only an attempt has been made to guess a possible reason of how the concept of White circuit came in the mind of the originator, but there are always other scopes for individual reader to guess under what circumstances such brilliant concept of resonant magnet configuration for magnet excitation was originated from 'Prof. White', which may reveal some other interesting facts on how the great minds work under pressure.

(b) Differences between r.f generator and White circuit schematics

Even though the function of two circuits are distinctly different, the similarities for the two schematic circuits are only drawn to suggest that the possibility of r.f.g design might have influenced Prof. White to imagine such a design of circuit and conceptualize a resonant mechanism to offer a suitable alternative solution to mechanical energy storage which was long due. The search concentrated to solve some critical problems such as how injected energy could be increased, how the energy from the magnet could be transferred with the new concept, where the energy could be stored, and how this energy could be retrieved before the next cycle of operation with minimum of power loss. Earlier, the feasibility to store huge quantity of energy either in inductive choke or in any capacitor was not seriously thought about or put into realization because of their functional, technical, and other limitations in retrieving energy within short intervals. As mentioned, it is purely a guess that this idea in dividing the magnet coil to parts might have originated from r.f.g design to reduce the quantum of energy stored in the magnet coil to a feasible value and easily transferred to an inductive storage choke and retrieved by using a resonant circuit mechanism. But, the problem of supplying makeup power for the energy loss still remained to be solved. In earlier accelerators, the synchronous motor was part of the rotating assembly and connected directly to the utility to draw the minimum power to make up for the losses incurred in rotating and stationary equipment. In this case also, similar arrangements had to be made to feed a.c supply of required frequency to the inductive choke which served as the transformer secondary coil.

(c) Comparison in power consumption for White and non-White circuit (ordinary circuit)

So far it has only been mentioned that the new concept of White circuit has reduced power consumption. In support of such statement, practical examples from operational results of different accelerator units are presented to justify and authenticate that claims. In comparison with the choice of "White circuit" (resonant magnet) against non-White circuit for magnet excitation, the first one is preferred because an excessive power is required for magnet excitation by the ordinary circuit (non-White circuit) in comparison to White circuit. This is best explained in (Glover 1994; Bothe n.d.; West et al. 2004), with RAL 800 MeV synchrotron, which showed when the peak power requirement for resonant system is 15.1 MVA only, such requirement for nonresonant ordinary circuit would have been 151 MVA, this huge difference in peak power requirement explains the advantage and preference for "White Circuit." Similar result can also be cited with other accelerators adopting the principle of White circuit. Experience in the calculation for energy consumption for BESSY-II (Germany) also revealed and established a wide difference in power consumption for the two schemes. The difference in direct dipole power consumption of 0.3 MW with White circuit makes wide difference against 4.0 MW peak for dipole powering without White circuit justifies the advantage in favor of White circuit. There are a number of similar examples with other accelerator projects to establish the huge difference in power consumption between White and non-White circuit.

(d) Selection of magnet inductance and choke inductance

In mechanical energy storage facilities described earlier the flywheel was designed to hold much higher energy than the actual requirements for magnet excitation. In inductive energy storage where the pulse rate is more than that of earlier synchrotrons, the energy stored in the choke is also kept higher in comparison to the energy in magnet inductance so that the disturbance due to fast energy swings is not reflected in the supply network. In such application where the resonant capacitor C_m

is fixed by the magnet inductance L_m, the capacitor C_{ch} is tied up with L_{ch} dictates the economics of operation based on choke/capacitance costs. The economics of energy storage in chokes with respect to capacitors governs the choice of choke inductance.

Thus, peak energy stored in choke is

$$\mathbf{W_{ch} = 1/2\,L_{ch}\left(I_{ch}\right)^2} \tag{4.9}$$

Assuming $\mathbf{L_{ch} = 2L_m}$ $\mathbf{(2\,L_m \le L_{ch} \le 3\,L_m)}$,

$$\mathbf{W_{ch} = 1/2 \times \left(2L_m\right)\left(3/4I_m\right)^2} \tag{4.10}$$

Considering the optimum ratio of magnet inductance to choke inductance to be 0.5, the ratio of stored energy in the choke W_{ch} and the stored energy in the magnet W_m maintains a relationship, $\frac{\boldsymbol{Wch}}{\boldsymbol{Wm}} = \mathbf{9/8}$. This selection of magnet inductance to choke inductance has been accepted in many projects and some of them are as shown in Table 4.3.

Distributed Resonant Cell Configuration

Cycling synchrotrons (1–50 hz) demands high voltage during the acceleration period for the magnetic field excitation. Energy transfer process in ring magnet with distributed capacitor banks and chokes proved to be most desirable, effective, and economical. In this case, the magnet voltage in each sector is distributed equally and the demand of peak magnetic energy is drastically reduced which in turn reduces

Table 4.3 Ratio $\frac{\boldsymbol{Wch}}{\boldsymbol{Wm}}$ in different projects

Project	Frequency (Hz)	Peak magnet energy (KJ)	Peak choke energy (KJ)	**Ratio** $\frac{\boldsymbol{Wch}}{\boldsymbol{Wm}}$
3 BeV Princeton	19	1300	1500	$\mathbf{\frac{9}{8}}$
4 GeV Daresbury	50	877	985	$\mathbf{\frac{9}{8}}$
8 GeV Farmilab	15	10.69	12.5	$\mathbf{\frac{9}{8}}$

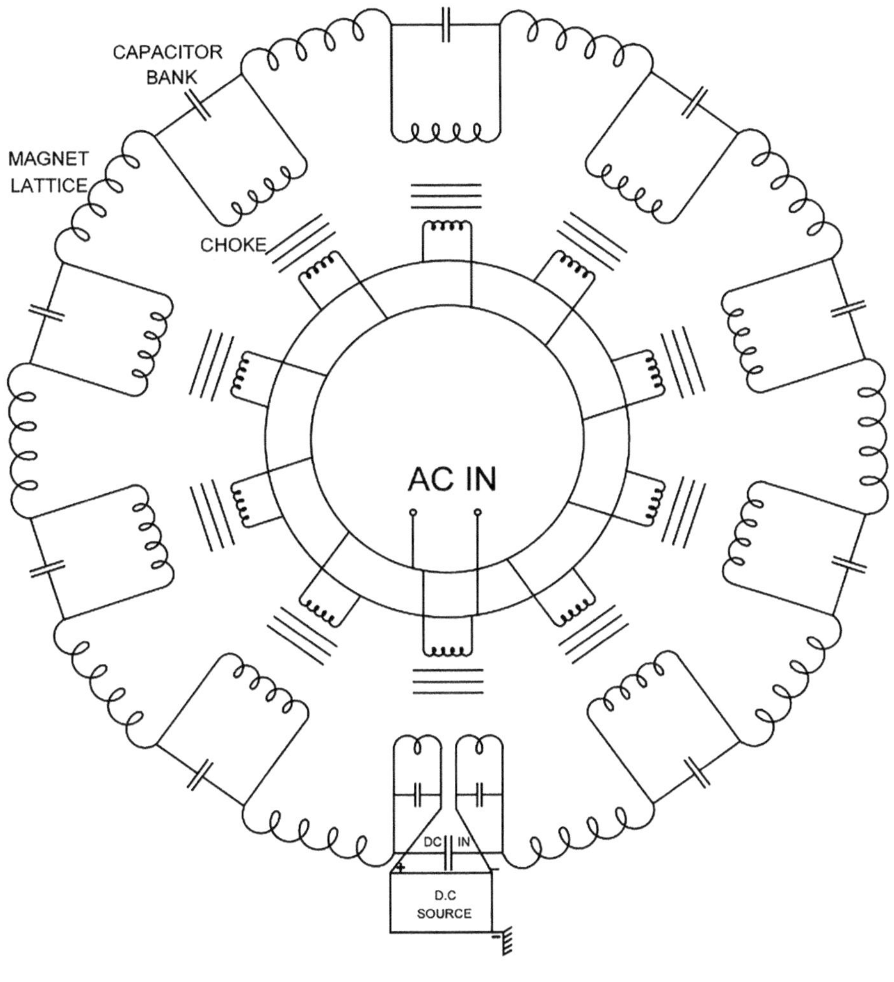

Fig. 4.10 Distribution cell structure-schematic

the value of the capacitor in a resonant circuit. Leaving aside the complexity of magnet physics involved with particle accelerators, the distributed magnet system is configured as segmented groups and the simplest structure consists of different types of magnet orientation for each group and their functions are to guide and steer the particle beam and confine them within the ring as per the design setup shown in Fig. 4.10. These magnets are composed of cells where each cell contains dipoles to bend the beam uniformly, quadrupoles for focusing the beam, sextupoles to correct chromaticity and other magnets for damping, compensation for distortions, control of imperfections, and guide the beam. In between the magnet groups, r.f stations for accelerating the beam are introduced as per the design demand. A simplified form developed for analyzing the functioning of multi-cell configuration is shown in Fig. 4.11 where each sector is connected in series to maintain current continuity.

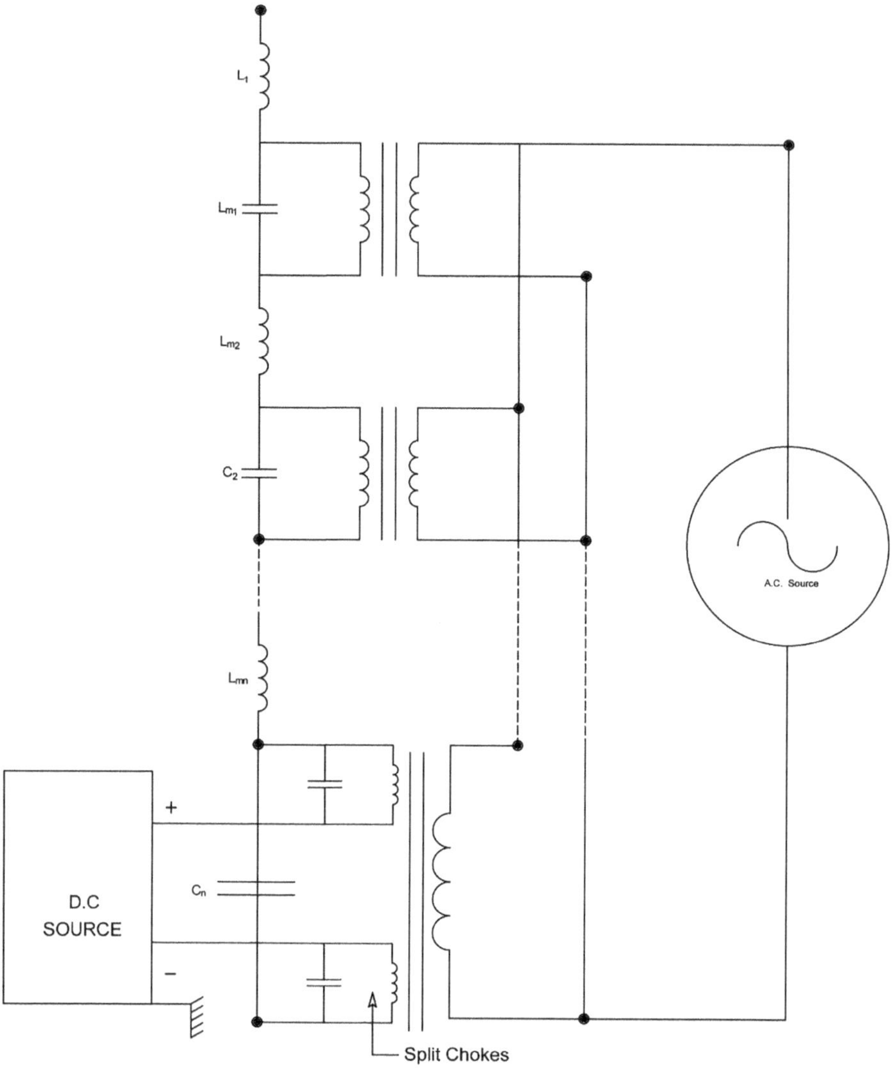

Fig. 4.11 Multi-cell configuration-simplified

The secondary chokes which are closely coupled to the parallel primary are split into n-sections to distribute the voltage equally across each. In this arrangement, one of the energy storage choke winding is split into two identical halves to supply d.c voltage and permit d.c bias current to flow in the circuit where one end of the split section is earthed and the maximum voltage to earth is 1/2n of total voltage. The limit set by the permissible operating voltages to ground of the system dictates the number of cells to be connected. The configuration of distributed resonant magnet system varies for different accelerators and it is shown how the number of magnet sectors is arranged to reach the required energy level in each projects. Under symmetrical conditions, the voltage of each magnet/capacitor group is zero, i.e.,

$$\mathbf{L_m\, di_m / dt + 1 / c \int \mathit{i_c} dt = 0} \tag{4.11}$$

The a.c ring voltage not being cumulative, as would be the case for a simple series connection of magnets, alternates around the ring. The a.c voltages across each choke winding and magnet group are identical in amplitude but reversed in phase and the d.c bias current is common to both. The prominent factors in the design depends on the operating frequency, pulse repetition rate, beam intensity, method of powering the dipoles and quadrupoles, dc and ac power supplies. The other features in the design include

- distribution of chokes and capacitors suitably in the available space
- fabricating a single massive choke with the idea of minimizing engineering problems in maintenance
- economic considerations; reducing the cost involved in the choice of components of the resonant excitation circuits and the number of r.f stations
- limiting the induced a.c voltage level in the system

The dynamic behavior of the circuit due to influence of mutual inductances of the magnetic coupling no longer remains simple with a large number of sectors but adds a complex problem in the real-time analysis of White circuit; therefore, its contribution to the electrical circuit cannot be ignored but properly analyzed. The distributed configurations emphasize on distinct periods for mathematical analysis and state equations are set for each transition period from one state to the other. This state space method simplifies the computations and provides insight into the physical phenomenon.

Application of Distributed Resonant Cell Configuration and White Circuit in RAL (Old RHEL)

The magnet excitation adopted in earlier design (mechanical energy storage) had two important hurdles; firstly, restoring the stored energy from the magnet coils during the decaying part and secondly, implementations of a suitable system for reducing the fluctuating load on the supply network. To get rid of motor–alternator and the massive flywheel set to reduce the fluctuation load on the power supply, the implementation of resonant method of excitation was found to be the only acceptable alternative solution at that time. The configuration adopted by "White circuit" eventually eliminated the rotating parts and resolved the fluctuating demand problems in much simpler way. The cumbersome arrangements with large rotating equipment along with the associated rectifier plant to supply power to the magnet coil had to be declared obsolete and abandoned with the new found technology. The huge cost of experimental setup such as Nimrod and other parallel accelerator projects mentioned earlier had to come to an end and terminated at some point of time. It is quite natural that in course of time the experimental need changes. To encourage and fulfill the aspirations and demand of the investigators, future research works are directed towards opening new frontiers of knowledge bearing in mind the greater benefit to be gained from such experiments with the cost involvement for the project. After the official declaration of closure of the project Nimrod in 1977, the British researchers decided to shift the focus of the experiment towards Spalled Neutron Source, subsequently renamed as ISIS facility, which was set up at the Rutherford Appleton Laboratory (Rutherford High Energy Laboratory renamed as RAL) in the dismantled site of Nimrod to install the new accelerator necessary for such work. Proposal to set up Spalled Neutron Source at the ISIS facility was taken up with a view to focus on the application potential of Neutron scattering and Muon spectroscopy in condensed matter and nuclear science. RAL was developed subsequently as the center for one of the world's brightest neutron source. So far the success of ISIS is the result of the intense pulse of neutrons that the facility provides for experimentation. After the decision to terminate Nimrod, the preparatory work for the new project site started with the aim to maximize the utilization of space and supporting equipment to reduce the project cost. Before taking any decision on new accelerator project, it is worth giving serious consideration in selecting suitable equipment and minimize the cost by restoring the old equipment from the dismantled site and make it useful for the required application as the availability of spares is difficult with the advancement of time and fast changing technology. Usually, the life cost of such projects need to be evaluated critically so that within the fixed project time span the future plans and the course of investigations can be changed in a direction well in advance to get the maximum benefit from the research. Thus equipment in any large accelerator facility which become obsolete and had to be discarded can be relocated at some other destination to reduce the cost of procurement. At this point, it will be interesting to know that BESSY-I (Berlin Electron Storage Ring Society for Synchrotron radiation), founded in 1979, at a cost of 66.5 Million Euro and operating at energy level of 200–500 MeV was dismantled in 1999 and its component parts were donated by German authorities to the SESAME project and shipped to Jordan. Subsequently, BESSY-I was replaced by BESSY-II (Berlin), 1.9 GeV, 10 Hz, a third-generation synchrotron radiation source in 1996 that produces extremely bright X-ray. Many such abandoned synchrotrons are either shifted to other places or relocated at the old site and utilized the discarded or unused components in the new project. It is difficult to pursue such scientific investigations with high energies in many countries due to excessive cost and restrictions on allocating fund for this type of project. In that case, hiring of facilities at other projects can also serve the research investigations. Such decisions are all the more vital and necessary for countries that have not taken any decision as yet to install an accelerator project. With the delay in the project time, not only the cost will increase, but the tuning of research topic is shifted by the time it comes into operation. For instance, when ISIS facility was developed at the old site of Nimrod, some of the equipment were salvaged from other accelerator installation, even the magnet coil used in NINA synchrotron of Daresbury Nuclear Physics Laboratory had to be shifted to the RAL site. The change in pattern of magnet excitation was an important deviation and implemented in future projects. At the heart of the ISIS accelerator, there is a large "energy storage choke Nina" to serve in the form of ring magnets with ten main windings connected in series with capacitor banks so that they form a resonant circuit and operate at the fundamental frequency of 50 Hz. Actually, when a heavy metal target is bombarded by firing short bursts of accelerated protons of high energy intense pulses of neutrons are produced. The process flow of proton beam is exactly the same that had earlier been mentioned. Only the method by which such high energy is imparted to the protons by synchrotron accelerator is different. In this case, the proton beams after the injection into the synchrotron are rotated around the ring and accelerated by r.f cavities, focused and guided by a series of magnets. The field frequency is precisely double that at which the protons rotate around the ring so that they get an accelerating kick every time they go past a cavity. It increases throughout the acceleration cycle to match the changing energies and rotational speeds of the protons and at the same time, the protons are squeezed into two bunches before being extracted and sent to the target station. The change in pattern of magnet excitation was an important deviation.

The change in pattern of magnet excitation was an important deviation and the decision to switch over to resonant excitation of the magnet has offered many advantages which encouraged other projects to implement the same. Actually, developing building facilities and commissioning a synchrotron accelerator is a very lengthy process and as mentioned earlier and repeated again that this time factor plays an important part in deciding the areas of research because by the time the system goes into operation, the focus on the dominant area for research work may be diverted to some other advanced field. So perhaps, in order to reduce the time as well as the cost of revamping, whatever could be salvaged from the remains of 7 GeV Proton Synchrotron was retained for proper utilization for the ISIS facility. Even, in many cases, equipment now being used are as old as 50 years. Mainly, the PLA which was operating at 15 MeV and the rest of the acceleration tanks that were in the pipe line during Nimrod operation as mentioned but, so far not taken into circuit, later upgraded with the addition of fourth tank to provide an injected energy of 70 MeV to increase the initial energy level at the entry to the torus. It was decided that the PLA, which was later upgraded to 70 MeV at the end phase of Nimrod, to be utilized as an injector to the new synchrotron project. At the time when ISIS Spallation neutron source produced its first beam on 16th December 1984, it used 665 KV Cockcroft-Walton pre-injector, a 70 MeV Linac and 800 MeV proton synchrotron with a cycling rate of 50 Hz. It was a new aspiration for researchers in the field of nuclear physics but, it was essentially a technological challenge to the engineers for the installation, operation, and maintenance of the highly sophisticated equipment with the latest art in design. It was expected that a higher degree of sophistication will be involved in future and to appreciate and understand the implication of the demand, qualified and best trained engineers are to be engaged for the job. With changes in decision, the electrical engineering configuration shifted drastically. Earlier model of electrical supply to the magnet was replaced with the design concept of the so-called White circuit.

Further Modification Plan at RAL

My last visit to Rutherford Appleton Laboratory (RAL) in 2009 was by courtesy Dr. David Findlay, head ISIS, who took keen interest in explaining the developments after the termination of Nimrod project. Thereafter, Mr. Mike Grover, Chief Engineer explained the technical details and showed me the project site, the arrangements for new setup under test for upgradation, shared the critical problems arising out of the electrical design, considering me as an old member of the family and discussed on the plans undertaken to handle the technical difficulties. My old memories were cropping up to adjust with the new arrangement of equipment. The synchrotron accelerator with "White Circuit" configuration used in RAL accelerator is shown in Fig. 4.12. In this case, the circuit was designed to resonate at the operating frequency of 50 Hz to reduce the reactive load on the makeup supply. It also had the advantage of ensuring that the same current flows in all magnets. Again, it could be seen how a d.c.supply was used to bias the sinusoidal magnet waveform. The d.c. bias was connected by splitting and center grounding one element of the White circuit whereas the a.c. makeup power was supplied to the primary winding of the energy storage choke. The advantage of introducing a d.c. bias was to increase the efficiency of the White circuit by reducing the total power requirement for the same change in magnetic field from injection to peak energy. A very important consideration in the design of "white circuit" was to maintain the resonant frequency. At the time of design, it was expected that the value of the capacitor may fluctuate due to change in temperature and accordingly in order to check the fluctuation of capacitor value, half of the capacitor was chosen with negative temperature coefficient and the other half with positive temperature coefficient. But, during operation it was found that such gradation of temperature was not required as the surrounding temperature could easily take care of maintaining the temperature equilibrium. Actually, the windings of the Nina choke were built into a single frame; therefore, the magnetic circuit was closely coupled as desired by the design for the best performance, but the drawback with such arrangement was that in the event of a failure of any winding, the choke would have to be dismantled to replace the damaged parts. The failure of any part of the coil could result in the shutdown for more than a year for rectification or replacement. The problem at present in the operation of synchrotron is to find out a solution so that the long downtime due to breakdown in any part of the coil could be minimized. Since the spares are becoming scarce, a new approach to replace the coil within a reasonable time is getting priority in the design. In the replacement model, the magnetic circuit is split into ten smaller chokes with main and auxiliary windings. The design of "white circuit" with modified choke is shown in Fig. 4.13. Earlier, it has been mentioned that the choice of a.c supply for continuous excitation can be from a direct a.c source or rotating machine. In this case a

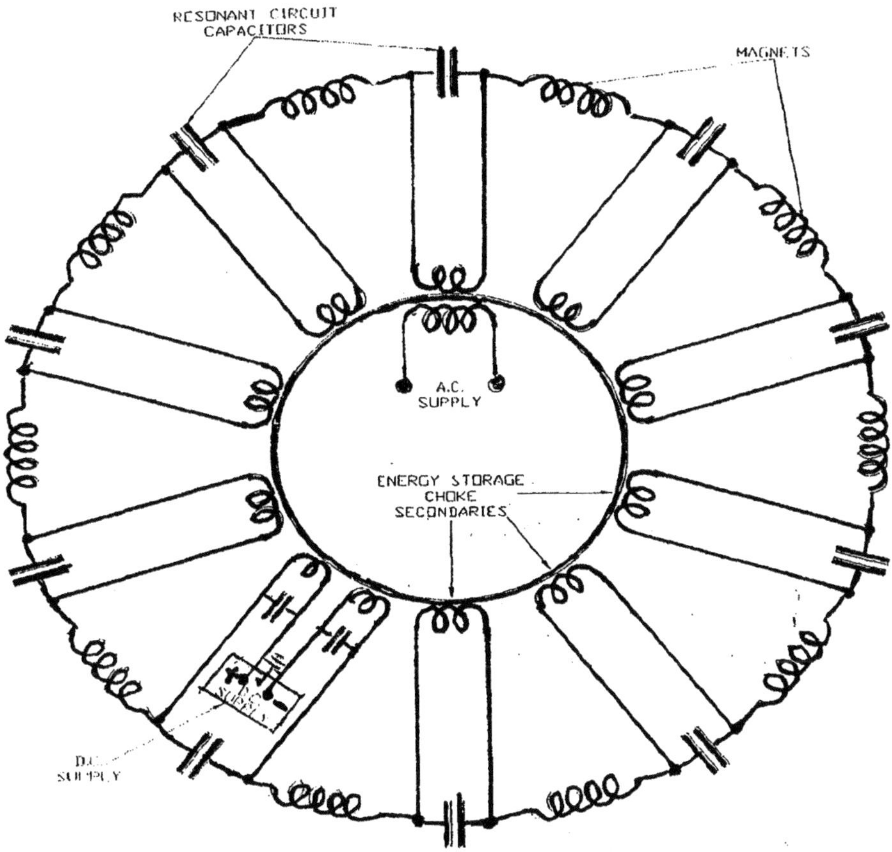

Fig. 4.12 ISIS ring magnet White circuit-Sync. Accl. Configuration in-RAL. Main P09

motor–alternator set is used to supply the a.c power, but the interesting part in the design is the conversion from 3-phase to single-phase a.c supply before feeding to energy storage choke. Figure 4.14 shows how the a.c is fed to the primary winding. Also, the processes of replacing the motor–alternator set with solid-state uninterrupted power supply sources were being considered and the alternative arrangements were being tested. Three numbers of uninterrupted power supply (UPS) sources were organized to supply the required power to the circuit. The upgrading of the a.c. supply to the main magnet was necessary to improve the system reliability, improve magnet current stability, and reduce the risk of mains failure. Distributed resonant magnet system vary for different accelerators as mentioned earlier but, primarily, all of them maintain a form very similar to ISIS synchrotron. Under symmetrical conditions, the voltage of each magnet/capacitor group is zero, i.e.,

$$\mathbf{L_m\, di_m / dt + 1 / c \int \mathit{i_c} dt = 0} \tag{4.12}$$

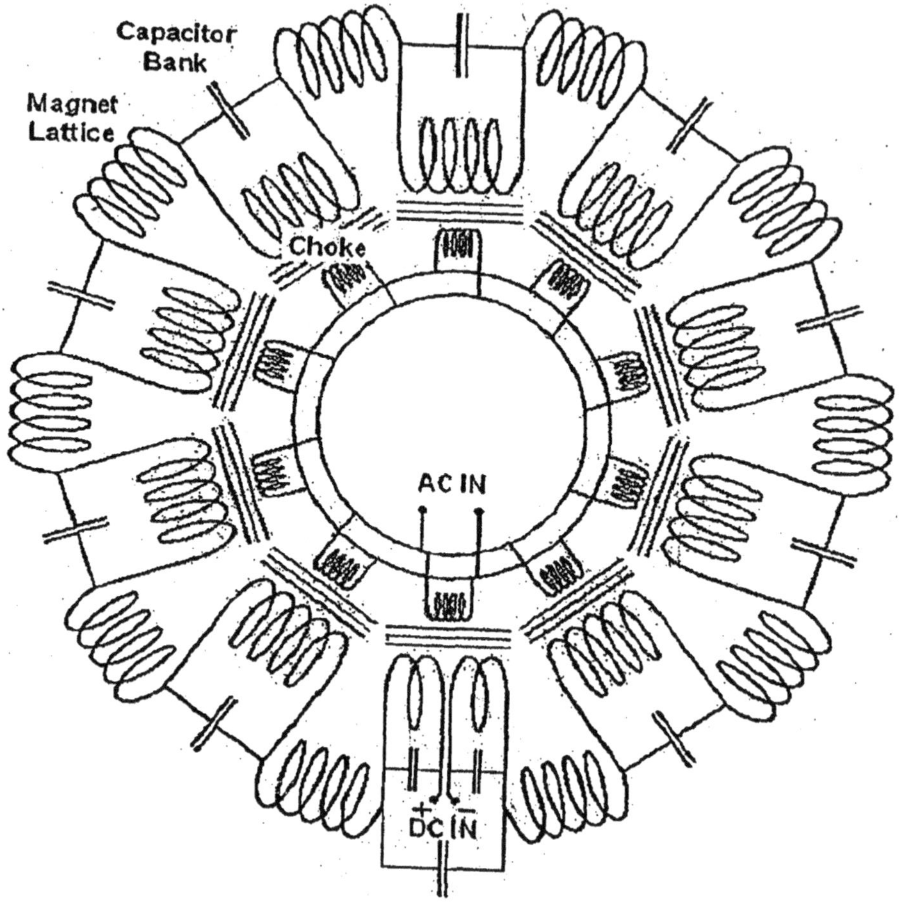

Fig. 4.13 White circuit with modified chokes at RAL. Main P11

The a.c ring voltage not being cumulative, as would be the case for a simple series connection of magnets, alternates around the ring. The a.c voltages across each choke winding and magnet group are identical in amplitude, but reversed in phase and the d.c bias current is common to both. The prominent factors in the design depends on the operating frequency, pulse repetition rate, beam intensity, method of powering the dipoles and quadrupoles, dc and ac power supplies.

Other Patterns of Magnet Excitations

It was a long-standing aspirations of the engineers to come out from the two most popular and dominant patterns of magnet excitation practiced between 1950 and 2000+ which served the purposes of scientific research at critical times despite

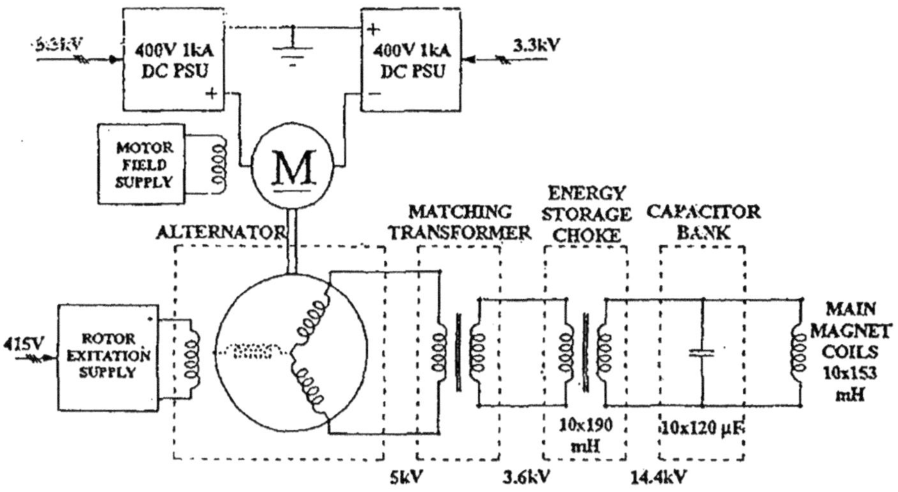

Fig. 4.14 Schematic of a.c supply (3/1ph) to inductive choke

several operational disadvantages. To find credible alternatives for changing the pattern of magnet excitations to give better performance other than the two mentioned, several projects were undertaken to find new methods. The rigorous search was a challenge to the specialists in the design of power supply to synchrotron accelerator with different forms of magnet excitation. Fascinating new possibilities emerged to support different patterns of magnet excitations with superior design concept to the facilities within the basic frame work of supply and demand of magnet current. The basic frame work implies to give stress on the following parameters:

- Supply – Pulse/Cycle (frequency)
- Demand – Controlled magnet load during
  - (i) acceleration
  - (ii) deceleration
- Constraints
  - (i) Maximum magnetic field is attained during acceleration.
  - (ii) Fast extraction of magnet energy during deceleration period.
  - (iii) Keep vibration of rotational components within limits.
  - (iv) Reduce power consumption.
  - (v) Limit harmonic generation.
  - (vi) Limit voltage fluctuation.
  - (vii) Rigid grid connection.

In order to generate new ideas, attempts were made to revive the schemes abandoned earlier and kept aside from the ambit of our evaluations for various reasons are brought forward for reassessment to find out their inherent shortcomings at the forefront of research. The priority list in the consideration for re-evaluation was how to utilize such schemes as change of patterns for magnet excitation. To narrow down the weaknesses of some abandoned pragrams such as direct supply from grid, discard flywheel from the complex rotational assembly, use of capacitors as feasible energy storage unit were revisited to analyze and explore how new design possibilities with the advance technology could be clubbed and fit in the operation to improve performance. The inspiration to take up such research on discarded systems is to study afresh the restrictions imposed on power supply arrangements for accelerator duties by the regulatory authorities and try out implementing new forms of technology support with upgraded devices. The attempts were made to relocate the storing of magnetic energy, exploit the functioning of new switching devices developed during the passage of time in the transfer of energy between supply side and load and focus on following aspects of the problems and search for solutions as follows:

(a) Direct supply from grid with imposed conditions
(b) Direct supply from grid to rotating equipment without flywheel
(c) Direct supply from grid to converter with capacitor as energy storage
(d) Energy transfer with fast-acting switching devices

Direct supply from the grid to converters with imposed conditions

The magnet power supplies on various projects differ on two important decisions as follows:

(i) Firstly, the effect of change on the grid supply to converters without any rotating parts as buffer to support the accelerator operation.
(ii) Secondly, the cause for concern is smaller time constant for the magnet load.

The feasibilities of implementing discarded schemes for magnet excitation with the support of new technology was very high for which continuous development and trial was initiated even when accelerator operations were active with two old systems. The past experience in handling motor generator sets including flywheel was major concern with pulse power loading in accelerator operation. On this issue, incidence of alternator failure at Rutherford Laboratory raised alarm on the choice of such system configuration which other accelerator complexes also supported this decision and subsequently declared genuine concern about the drawbacks in adopting such scheme. With the negative report from all user facilities such scheme of magnet excitations faced serious concern to continue and researchers initiated alternative approach for solving the problem differently. But, all these attempts opened up exciting schemes where the chances of utilizing the operational experience of the past in every move was an added advantage. The modifications were taken up for the development of these new schemes with the support of modern technology and long working experience on multidisciplinary subjects and compare their performances with the past systems to assess the improvement. The steps taken for these new ventures were risky but handled with caution and confidence. The possibility of grid supply directly to converters for magnet excitation and return the stored energy back during the deceleration period was included in the evaluation earlier, but it deterred us at that time from accepting such scheme of magnet excitation due to the weakness of the available grid in accepting fluctuating voltages. It was well understood that the supply to the magnets of a large accelerator poses few problems when the power involved is considerably higher since the current has to be pulsed very precisely to give the required field variation. Apart from that, the quantum of harmonics generated in converter units due to pulse loading and their effects on the neighboring consumers required provision of some form of ripple filter network for minimizing harmonic generation to tolerable level. The suggestion to install adequate compensators at the bus bar end to reduce reactive power consumption and improve power factor was also an important part of such design. The accelerator facilities might have taken the decision that the question of pulse loading and harmonic generation with direct supply need to be re-evaluated to separate out the problem and solve individually to make the new system design viable and acceptable for accelerator operation. CERN Geneva and other research organizations with their robust technological background initiated extensive studies for a long period to find some better alternatives for magnet excitations. The 25 GeV proton synchrotron developed in November 1959 and the operation of which had been compared very often with RHEL and other accelerators of the time was later upgraded to 26 GeV. CERN's upgraded proton synchrotron (PS) started transferring high energy particles to 300 GeV Super Proton Synchrotron designed to function as an intermediate stage between proton synchrotron (PS) and LHC to produce much higher energies for the benefit of scientific studies. This was also upgraded in stages later to generate 450 GeV. The development of such high energy machine required huge funding and this was only possible with the cooperation of the member states jointly agreeing to share the fund and made the accelerator operational in 1976. SPS (Super Proton Synchrotron) acted as the trial ground and undergone many alterations and modifications with new innovative technologies to increase the intensity of the proton beam. The support of new technologies in the development process is required to achieve very high precision depending on the size, the energy level of accelerator, the scale of performance improvement, and reduction in operating cost. This need of precision for the parameters is required with the energy level of the synchrotron; otherwise, the protons will be lost before they had completed acceleration to their full energy. This is especially true in case of SPS where it is necessary to keep the magnetic properties of the individual bending and focusing magnets within 10^4 ppm. It was realized that proton energies of over 300 GeV could be attained by taking the advantage of improved techniques and using separate dipole and quadrupole magnets for bending and focusing which allowed a higher effective bending fields than was possible with earlier combined function of magnets. All the research organizations working on higher energies are seriously addressing the lacunaes which prevented them to adopt the earlier schemes in accelerator supply and find out how the new modifications can be effectively included in the system. Mostly, the research format by CERN and other organizations concentrated to work on three possible directions:

(i) Utilization of direct supply from grid
(ii) Minimize rotating equipment

(iii) Utilize capacitors as storage element

The implementation of direct supply for magnet excitation to earlier accelerators was stalled mainly due to power supply system not being adequately equipped to provide very rigid and high voltage grid with large short circuit capacity. Actually, the possibility of pulsing the magnet directly from the high voltage grid is serious concern; therefore, all loopholes which degrade the performance required to be plugged before approving the facilities to initiate operation. The task for new design is to ensure removing of difficulties and take adequate precautions to guard the system from failures. Considerable amount in maintenance expenses and a sizeable fraction of the total cost is to be included to maintain the operational reliability of the accelerators without affecting the neighboring consumers of such system. The moot point during the evaluation of direct or any other type of supply to magnet can only be assessed on the quality of performance improvement and the cost. In other words, the available grid which is not capable of withstanding high energy swing originated during the course of magnet excitation demands further studies. For instance, the grid voltage of 33 KV at the common coupling point at RHEL (Rutherford High Energy Laboratory), UK, was not equipped to supply power to NIMROD directly for magnet excitation, and this was also true for all earlier accelerators. Even, power supply for 25 GeV proton synchrotron at CERN in 1959 was not directly connected to the grid at the initial stage of operation due to lack of robust power supply with high short circuit level at the point of common coupling. However, the situation changed later when this was modified to 26 GeV machine to generate higher energies for SPS, Super Proton Synchrotron, and connected to rigid grid supply. The shape of the current, voltage, and the active power to the magnet of SPS were powered from dedicated power systems and varied at the magnet terminals from ±40 MW. When such was the case, engineers from CERN possibly deviated from earlier decisions of abandoning direct supply but stepped up their search to alter the design philosophy to implement direct supply differently and justify its applicability as a part of the challenge for an alternative.

On the other side, failures in rotating machines due to pulse loading and the consequences of such mechanical failures either partly or fully, affected the accelerator operation even after the number of precautions being taken in all fronts. The motor–alternator sets designed to stand as a buffer between the supply and load became the potential source of trouble due to periodic torque reversals. The decision to discard the flywheel from the rotating setup which was considered to be an essential part of the complex rotating equipment as storage of magnet energy against all odds at that time was under the scanner to handle such problems in two different ways as follows:

(i) The first one was to reassess the possibilities of direct supply.
(ii) The second one was to discard the flywheel from the setup of rotating equipment without disturbing the scheme of motor–generator assembly.

The solution to first technique is applicable when supply is connected to converter from robust grid but again, supplying network power directly to the converter is only possible when the grid could be strengthened as per the guidelines of the supply authorities such as 400 KV European grid connections with 12 GVA short circuit level stands as rigid support for supplying power directly to converter of 1 HZ, 60 MW load of SPS, CERN accelerator magnet from 18 KV PCC (Hettel et al. 1991; CERN 2011). In the case of SPS, the peak power during the field rise is 135 MW and the total stored energy at full field is 82 MJ. In order to bring the field down to injection level rapidly for the next pulse, the peak rate of withdrawal of this energy needs to be very high. In the past, the experience of failures occurred in rotating machines due to mechanical stresses developed during field reversals created various operational problems. Although there have been some breakdowns with SPS, nevertheless the accelerator has been working successfully for 75% of the scheduled time. In November 1967, RHEL engineers also conducted similar tests to explore the consequences on e.h.v transmission network supporting pcc (point of common coupling) for supplying power directly to the converters for a large pulsed load (Fahrini 2007). The purpose of such exercise was to check the impact of system performance at two different voltages: 132 KV and 275 KV and at different load levels under pulsating load conditions. The result of the tests showed that the pulse power propagated into the 400 KV super grid produced little or no attenuation in the local 132 KV network. The large growth rate of the installed capacity of public power systems and the advent of e.h.v transmission network suggested that a direct connection of the pulse load to converter is possible when a uniform distribution of the pulse energy is achieved (Dwek et al. 1970). But the possibility of supplying a large pulsating load directly from the national grid requires thorough examinations to satisfy few imposed conditions mentioned below before it is acceptable to the supply authorities.

(i) Voltage fluctuation must be minimized at the point of common coupling.
(ii) Load must be substantially free from harmonics.
(iii) The system frequency disturbances must be within limits.

Actually, in these types of connections, special emphasis is given on the factors responsible for variation of voltage and frequency of the supply mains due to magnet load. It is usual to provide adequate compensators as Var generating system at the bus bars to clamp down the stresses on the grid feeding the magnets of CERN or any other accelerator. The thyristor control converter circuits always consume reactive power; therefore, arrangements of suitable compensators are necessary at the bus bar terminal to improve power factor, reduce reactive power consumption, and satisfy the neighboring consumers as per the regulation of the supply authorities. In order to reduce voltage fluctuation, the reactive power compensation must be a function of the reactive and the active power consumption of the rectifiers. There are various possibilities of compensation such as:

I) Saturable reactors combined with capacitor bank
II) Capacitors switched on and off stepwise by thyristors

The choice will depend mainly on their respective costs. Usually, filter chokes are connected in series with the capacitors used for reactive compensation in order to suppress the harmonic voltage distortions produced by rectifiers. In industrial

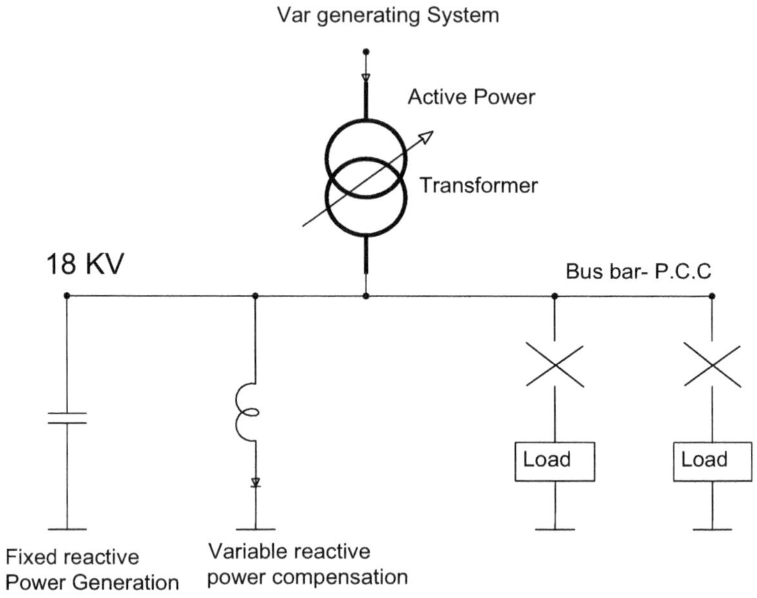

Fig. 4.15 Schematic of reactive power compensation

applications, this is quite normal and to avoid the disturbances to the neighboring consumers and improve the overall power factor, synchronous loads are segregated and methodically distributed in the bus bars to raise the power factor. Even in arc furnace, load arrangements are also made to connect capacitors in the form of fixed reactive power generators or introduce variable reactive power compensators. The use of saturated reactors are now being replaced by high voltage high current thyristors connected to the capacitor bridge compensator and considered as excellent means of continuous control of Vars. Schematic of reactive power compensation is shown in Fig.4.15. The fact is that static arrangement of controlling saturable reactors and capacitors could have been of advantage for Nimrod operation also, where the fluctuation produced in the supply network owing to the operation was limited to 10 MVA. But, the scheme of such compensation at the bus bar was possibly not adequate due to nonavailability of suitable components at that time; however, the installation would have been very costly. Actually, the problem was anticipated when trying to obtain very precise pulse-to-pulse repeatability, particularly during injection times. SVC with direct supply and capacitive energy storage is shown in Fig. 4.16.

The implementation of the second technique is crucial and only possible when the function of the flywheel with such rotating mass can be compensated by choosing an alternative solution for storing the huge magnet energy. As the large active power swing (±40 MW) in SPS during a cycle mentioned earlier was not acceptable to the electrical network, the thrust of kinetic energy storage required to be transferred to motor–generator assembly. In such a situation, the solution could be that

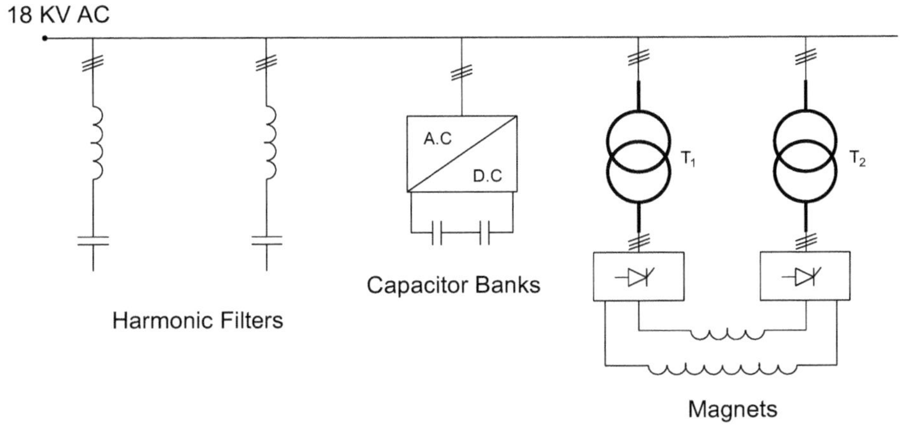

Fig. 4.16 CERN SVC with capacitive energy storage

sufficient mass equivalent of flywheel need to be transferred, distributed, and shared by the alternator to store the massive energy of the magnet. In order to execute the plan of such transfer, specifically designed motor–alternator assembly was ultimately considered to be an acceptable solution for storing the magnet energy. CERN possibly implemented withdrawal of flywheel, a rotational kinetic energy component in the system and adopted the technique to compensate by raising additional weight equivalent of flywheel to alternator to store the energy of the magnet. But, such extraordinary mass of the rotor required special technical skill in designing and manufacturing. Accelerator operation for magnet excitation based on modified kinetic energy storage started from 1968. The 80 ton rotor of the 95 MVA alternator drives a 6.6 KV network feeding four 12 MVA transformer of the rectifiers which power the magnets by delivering d.c.current of 5.5 KA at a voltage of 9 KV***. The alternator which stores kinetic energy of 233 MJ during the cycle is allowed to drop its speed from 1020 rpm to 980 rpm. Successful replacement of flywheel was achieved mainly when the manufacturer cooperated fully to meet the challenges by specifically designing alternator to serve the purpose. This again confirms the amount of cooperation require for the operation of accelerator from the outside agencies while treading in unknown territories. AGS booster synchrotron at Brookhaven National Lab, New York which operates within 0.3–7.5 Hz also takes the energy for the machine directly from the grid and the energy is stored in a rotating machine. The results of the two techniques are considered to be most significant as viable alternatives to other two prominent patterns of power supply for magnet excitations that continued and dominated in earlier accelerator operation. Another most significant development for magnet excitation at present is the growing application of direct supply of mains utilizing capacitive energy storage system for the magnet energy. With the efforts made, magnet excitations remained no longer limited to intermittent excitation from pulse power supply with mechanical energy storage or continuous excitation of White circuit resonant magnet with inductive energy storage for accelerators but initiated once more the change of patterns for magnet excitation schemes with the list as in Table 4.5.

Capacitive Energy Storage with Switch Mode Converter

The study for alternative methods of magnet excitation based on capacitive energy storage started when the availability of required capacitor to withstand desired duty became a reality. Actually, the preference for capacitor was always high as an ideal storage element for electrical and electronic circuits functioning at higher frequencies. Even then, the selection of capacitor for energy storage duty in accelerator operation was not accepted because the application of capacitors was found to be unsuitable as energy storage device for magnet excitation due to the number of reasons such as difficulties in handling, limited energy density for storing high magnitude of magnet power, inability to withstand severe duty cycle which will affect proper functioning where long time constant is required and nonavailability of superior quality capacitors for a long time which could store more energy per unit volume to support such storage. Apart from all those, the capacitor being a temperature-sensitive element, effective temperature monitoring and control is required for resonant circuit operation and as a result of these deficiencies its selection was critical and kept in abeyance for a long period for accelerator duties. So, all efforts were diverted towards developing suitable capacitor technology which could

Table 4.5 Change of patterns for magnet excitation over the years

Sr. no.	Supply type	Storage type	Scheme	Projects
1.	Pulsed power	Mechanical energy storage	Motor–flywheel–alternator–transformer–converter–magnet	Synchrotrons of 1950s
2.	A.C Cycling frequency 1–50 Hz	Inductive energy storage	White circuit; resonant magnet	3 GeV Princeton, / NINA A.C to inductive choke
3.	Direct A.C supply (Rigid grid)	Capacitive energy storage	SPS type CERN	A.C to converter
4.	Direct A.C supply (Rigid grid)	Mechanical energy storage	Motor–alternator converter–magnet (Flywheel discarded)	CERN
5.	A.C Cycling frequency 1–50 Hz	Capacitive energy storage	Switched mode converter	SLS, CLS, Diamond, and others Fixed voltage d.c; PWM, IGBT
6.	A.C/D.C Converter	Capacitor energy storage	CERN type	CERN type

give a better solution in accelerator operation. The continued research resulted in the availability of high voltage superior quality capacitor suitable for industrial application and capable of withstanding a large number of charging–discharging duties with virtually no burden of maintenance. Therefore, before considering capacitors as a storage device, its drawbacks and suitability for particle acceleration duties are to be examined critically. The development program of new type of capacitors was initiated along the same time with the inductive energy storage schemes and in parallel line to satisfy the requirements of accelerator operation. The field testing of capacitors with new found alternative technology resulted in declaring capacitors as a preferred choice for storage device and initiated promptly to take the advantage of changing the pattern of power supply for magnet excitation. The configuration of capacitive energy storage with the innovative switch stands as buffer and a support between the magnet load, and the supply side during each cycle in place of inductive storage choke. Earlier, it was mentioned that the mechanical switch is commonly used in industrial application in d.c drive for regenerative braking applications. But, such mechanical change over switch cannot function at higher frequencies hence discarded from the evaluation process. With new found semiconductor device IGBT, it has become possible to develop switched mode converter topology, envisaged to be far better alternative to mechanical switch and can function with minimum breakdowns without inviting any hazards. The development and introduction of IGBT (insulated gate bipolar transistor) to function as the preferred switch has enhanced the popularity of the scheme. Its application to high speed operation with magnet system started playing a key role as buffer between the magnet circuit and the supply side to offer improved performance with electrolytic capacitors of suitable size. The acceptance of switched mode power converter with suitable capacitor bank designed by PSI (Paul Scherrer Institute, Switzerland) is considered as the pioneer and installed in various projects as a change in the pattern for magnet excitation system. Since then, the availability of high voltage electrolytic capacitors has contributed enormously in designing customized switched mode converter controls for accelerator operations utilizing high power IGBT. The design of capacitive energy storage with switched mode converter supply have gained sufficient interest and established its popularity very fast. Apart from that, CERN model of power converter shown in Fig. 4.17 with capacitor banks in the application of magnet excitation supply for synchrotron accelerators are new techniques in the series and surfacing as prominent with separate methodologies to replace the previous patterns of popular power supplies and implemented in few accelerator facilities. In this configuration, the stored energy in the capacitors are exchanged with the magnets during each cycle via d.c–d.c converter banks connected to the magnets. Only two a.c–d.c converters are connected to the mains and supplies the losses. The results on long-term performance and economic evaluations of the new schemes will be a milestone for the new projects to follow. These are some of the "changing patterns of magnet excitation" and diversion from the initial practice of accelerating particles which continued from 1950s.

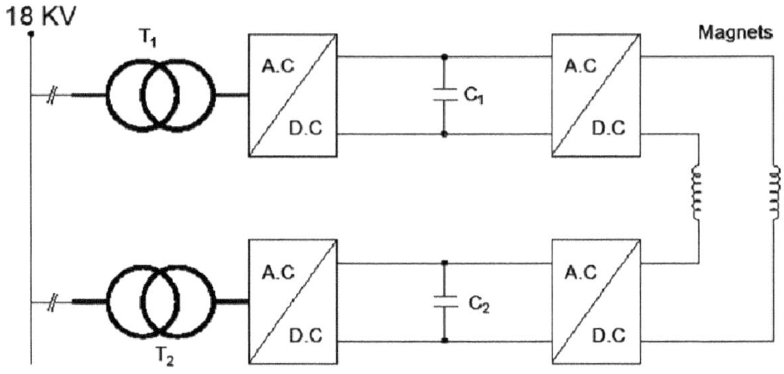

Fig. 4.17 Power converter with capacitor bank model with capacitor bank

Summing Up

Although flywheels in mechanical energy storage have long operational life, high power density for environmental impact and store energy with no upper limit yet, the technology of inductive and capacitive storage found to be economical and superior in most respects for particle accelerators. Upgradation of accelerator facilities is considered as a continuous process. Each time a new accelerator project is taken up, the question of right choice of magnet excitation, cycle frequency, and corresponding technology of power supplies are seriously evaluated at the initial stage before the design is frozen. The reasons for the changing patterns of magnet excitations of synchrotrons have been elaborated earlier. Upgradation of beam energies, injection power, rotating machinery if any, a.c and d.c power supplies to achieve high stability with the improved circuit design and adopting converter topologies with the latest state of the art in electronics, improvement of inductive energy storage chokes with higher grade silicon steel laminations, capacitor selection considering the temperature variations and replacements with corrective measures, modifications in the design of r.f are only a few in the list of improving performance. Quality of instrumentation and control with regard to current stability and limit of ripple factors are important and as regards providing better accelerator facilities there are many factors other than raw beam power that determine the success or failure of facilities, such as reliability, cost of investment, provision of technical support, and engagement of the user community with the facilities being offered, makes all the difference.

References

Bothe W (n.d.) Resonant excitation of synchrotron accelerators. Rutherford Appleton Laboratory, Chilton, Didcot, UK

Bouteille JF (2015) Power converter for cycling machines, ESRF
CERN (2011) Current state of CERN's High Voltage Grid, Engineering dept. Technical note, Ref. ELG-EVHT-ON-0001
Dwek MG et al (1970) Direct supply of a large pulsed load from e.h.v transmission network. Proc. IEE:998
Fahrini C (2007) "A novel 60 MW power system based on capacitor energy storage" for particle accelerator- CERN
Fox JA (1965) Resonant magnet network and power supply for the 4 GeV electron synchrotron NINA. Proc. IEE 112(6):1107–1126
Glover M et al, Digital control system for the ISIS synchrotron magnet power supply, Fourth European Particle Accelerator Conference, London, June 27-July 1, 1994, P-1830-1832
Hettel R et al, (1991) The 10Hz Resonant Magnet Power Supply for the SSRL, 3 GeV Injector, IEEE Particle Accelerator conference, Conference record June 1991, P-926-928
West S et al, (2004) Upgrade of the ISIS main power supply, Rutherford Appleton Laboratory, Chilton, Didcot, UK
White MG et al. 3 BeV High intensity proton synchrotron, proceedings of CERN symposium, June, 1956

Chapter 5
Semiconductor Devices, Converter Topologies, and Control System Strategies for Accelerators

Introduction

The selection of suitable semiconductor devices, correct choice of converter topologies—a generic method of power conversion and synthesis, along with robust control strategies are a priory condition for the trio to form a matching package to meet the stringent demand of magnet physics. Without this trinity working in synchronism, the purpose of accelerator control will be defeated. The earlier practice of selecting Hg–arc rectifiers contributed substantial losses during switching and conducting in converters to bring down the efficiency and performance of the synchrotron operation. The design of distribution cell structure based on the White circuit principle for magnet excitation also demands strict adherence to the supply requirements for efficient functioning of synchrotrons. It is therefore essential to upgrade converter topologies and adopt advanced control system strategies for controlling parameter tolerances to produce more protons per pulse and at the same time improve accelerator operation. Due to fast availability of semiconductor devices in various forms with different power ratings and circuit configurations, it is now possible to implement latest technology in the design of power supplies and control systems in present-day synchrotrons.

Semiconductor Devices

Effectiveness of a converter technology for synchrotron operation depends on the knowledge and the usefulness of essential tools and tackles, i.e., the availability and the selection of devices suitable to function as power converters with their limitations. The era of mechanical energy storage for magnet excitation in accelerator has

P. K. Dey, *Power Supplies for Synchrotron Accelerators*, Springer Tracts in Electrical and Electronics Engineering,
https://doi.org/10.1007/978-981-97-7606-1_5

been declared obsolete long back, but it taught us enormous lessons in understanding the type of services required for the operation of synchrotrons, connection of transformers and rectifiers in various forms for minimizing harmonics, selection of pulse firing circuits, techniques of minimizing ripple content, limitations of converter configurations, its drawbacks, and its scope for performance improvements with the advanced technology. Prior to 1960s, it was the era of vacuum tubes and power supplies or electronic circuits during that period mainly comprised of electronic valves. But, from early-1960s, a transition phase started to replace vacuum tubes with semiconductor devices. I had the opportunity to witness the impact of such drastic change in production while working with M.O. Valve Company in London for a short duration. The company was reputed for producing different types of vacuum tubes and the demand of its products had huge global market. It was impressive to see the product manufacturing stage followed by the functioning of quality control department where, the testing section engaged sincerely in implementing laid down procedures to maintain required standards and stringent specification demand of parameters set out for each type of valve systematically and rigorously in testing bay. The production of valves started slowing down from mid 1960s and ultimately gave way to the demand of new technology. Semiconductor diodes and transistors, the promising devices, gradually replaced electronic valves and metal oxide rectifiers. Those were the days of the past technology.

Thyristor (SCR), being a positive feedback semiconductor device, made a significant breakthrough in late 1950s as a gate control switch, which was quickly accepted as a suitable replacement for Thyratrons and mercury–arc rectifiers for its various advantages. To a large extent, it is the small size of the thyristor and its low switching losses in comparison to Hg–arc rectifiers established its effectiveness during operation within a very short period and made it acceptable to the user community as an alternative to other form of switching device. Only consideration was that, thyristor being susceptible to overloads, needed extra protection against excess voltage when non-conducting and excess current when conducting. The popularity of semiconductor devices demanded the power handling capacities to be raised gradually with the improvements of converter configurations and testing its suitability, selection of proper fuses for protecting the costly devices, providing effective cooling arrangements for limiting temperature rise for powering the magnets. "Power Electronics" grown slowly as a multidisciplinary subject capable of maintaining a balance between the supply side and the demand side by accepting the challenges of power supply designs suitable for the crucial magnet loads in the operation of particle accelerators with regard to accuracy, reliability, and stability of current with improved performances. Real challenge to SCR came after the emergence of self-commutating devices, suitable for high power and high frequency applications. Some notable of them are Gate Turn-Off switches (GTO-1960s), a positive feedback device like thyristor, power Darlington transistors (BJT-1975), a bipolar device with switching speed considerably faster than thyristors, whose current gain is enhanced with power Darlington configuration. The subsequent development of power MOSFETs (1978) and Insulated Gate Bipolar Transistors (IGBTs-1983) have made significant changes in applications. Important features of MOSFETs are its rugged switching performance and increase reliability, whereas the structure of an IGBT is more like a thyristor and MOSFET, but its voltage drop is lower than MOSFET of similar ratings. The formation of IGBT is a MOSFET coupled to a bipolar transistor and considered as a pseudo Darlington, or in other words a MOSFET as the driver and a bipolar transistor as the power stage. SCRs being costly devices are to be protected from over current against faults, commonly occurring in converters and bridge networks at the instant of short circuits, with high speed HRC fuses. I^2t fuses applicable in these cases are the measure of heat capacity of the semiconductor element which varies with the applied voltage and peak value of the short circuit current. The fuse rating for protecting SCRs is kept lower than that of I^2t value, whereas the snubber networks protect the devices from excess di/dt and dv/dt. The advantage of GTO, a power switching device having I^2t withstanding capability similar to SCRs is also protected by semiconductor fuses. On the other side, power transistors like BJTs and MOSFETs are not protected by semiconductor fuses, as they do not have I^2t withstanding capabilities and for protecting these devices some other electronic techniques are developed. The choices of converter and inverter grade thyristors need to be assessed properly because their turn-off times are distinctly different which few microseconds lower for inverter besides, inverter grade thyristors are turned off by forcing the current zero using external commutation circuits. In order to maintain power quality and strict adherence to such demand, the converter configurations for power conversion are designed to follow certain guidelines imposed by the supply authorities. After receiving the demand requisition from the magnet physics and accelerator section, the quantum of load connected for powering the magnets is assessed by the power electronics division where emphasis is given on calculating the total power supply requirements for r.f. generators and magnet loads for bending, focusing, steering, and course correction for the booster and storage ring with due importance to the allowable tolerances for producing the required beam for the charge particle before selecting a suitable converter configuration from the available devices.

Converter Topologies

Selecting converter topologies are a serious design challenge for the power supplies to particle accelerators. Magnet operation has a strong impact on the power converter topologies and on the ratings of power converter devices because nearly 70–90% of the total energy is consumed by the magnet system, r.f. generators, and beam line magnets. Hence, suitable converter topologies are required to be developed as per the demand from the available semiconductor devices with their inherent limitations. Different requirements, different applications, different type of rampings, and different cycling frequencies make the design of accelerator magnets and power converters to proceed in parallel. Alternative technology developed by Prof. M.G. White, specifically adopts resonant magnet excitations known as "White Circuit," where magnet current is represented by

$$I_{m} = I_{d.c.} - I_{a.c.} \cos(2\pi ft) \quad (5.1)$$

where

I_m = magnet current
$I_{d.c.}$ = d.c. current
$I_{a.c.}$ = a.c. current
f = cycle frequency

There are some properties of converter configurations which are unaffected by changes of shape, size, characteristics of devices but require concepts essential for the understanding of power converter topologies relevant in its operation to give the best performance. The development of power semiconductor devices, improvement in voltage and current ratings, higher power densities, effective protective devices, upgraded product quality, all added up to contribute in optimization of topologies for the advancement of power converters. Selection of the type of converter topology is guided by the parameters set out in the demand requisition; the level of voltages, the minimum and maximum current requirements, the ramp rates, operating frequency of the accelerator, and other factors. Any changes of parameters at the initial stage of design to reduce the cost component of the project are a compromise. In this regard, the effect of the operating cycle frequency requirements on power converters for cycling machines has been described (Bouterille 2015; Burnet 2016; Marks 2004). The power converters are vital parts of a cycling accelerator and as mentioned earlier the essential parts of the design is to improve the operational efficiency, control parameter drifts, and maintain stability and accuracy to the highest level. The reason for such caution for converter design will be further discussed later. The choice of power converter system for 3 GeV booster synchrotron (Marks 1996) also gives some guidance.

The basic structure of converter topology stands on two important demands:

1. The selection of suitable semiconductors for supplying d.c. to r.f. circuits
2. Generating current and voltage waveform to supply d.c. and low frequency a.c. for magnet excitation

The powering for the lattice magnets and r.f. generators are two essential requirements for the accelerator operation for which the selection of a cost-effective magnet excitation system is important after evaluating all the available schemes of energy transfer and storage possibilities such as mechanical, inductive, capacitive, and direct supply. The criterion for selecting direct supply from grid to power the converters for magnet excitation is separate and need to be assessed as per the criteria mentioned earlier. The loads of different magnets are required to be calculated after finding out the accelerator or combination of accelerators with their orientation, characteristics, and allowable tolerances to maintain the course of beam dynamics to operate at specified energy levels before the selection of power converters. Power converters for cycling accelerators are designed to ensure that it is capable of transferring/storing the energy which oscillates between lattice magnets and converter during the deceleration period. In booster synchrotron, the direct current is held accurately constant and a.c. is mainly supplied and controlled in such a way that adherence to the time interval is strictly maintained during which the field in the magnet is reversed. As the name implies the storage ring synchrotron specifically stores the charged particles after the particles are transferred from the booster to synchrotron; therefore, increase in energy in the storage ring magnet is slow. The momentum of the particle at the storage ring is either fixed or varies slowly with time for which the constant current supply from d.c. source is considered adequate for the purpose. In all cases, supply constraints and the protection of devices are taken up while framing up the converter design for supplies to the magnets. Many transformer topologies exist for primary and secondary side connections to rectifier configurations for different energy capacity of accelerators. The contributions of transformer connections and rectifier topologies referred earlier in the operation of synchrotrons of 1950s, including the 24 phase connections to rectifiers with ripple reduction technique of RHEL (Rutherford High Energy Laboratory), opens up wide scope for analysis and adoption of such schemes with upgraded products are likely to influence future developments. Economic design of transformers and the choice of a topology over the other involve a comparison between the ease of supply connection, selection of transformers, choice of rectifiers; mainly semiconductors, construction of the units, overall performance of the package, and above all the cost component for the project. The converter design is a compromise between the specified performance requirement and the restrictions on the a.c. supply network. The quality of supply that would affect the neighboring network impose limitations on the users and any such objectionable factors need to be addressed at the initial stage so that, the choice of a particular type of converter topology is free from the hassles of the supply authorities. The choice of converter topology depends on certain factors such as

1. The choice of transformer primary connections at the supply side and secondary connections to converter bridges
2. The selection of semiconductor devices, bridge network for fixed voltage d.c. or variable voltage d.c. output
3. Type of pulse firing circuits to be selected
4. Type of fault protective methods to be employed
5. Choice of ripple filters

In the operation of a converter circuit, all the above factors are usually guided by the manufacture's manuals, available technical literatures, and from the knowledge and experience of experts in the field in protecting the devices from unwarranted faults during operation. The internal disturbances reflected to the supply side such as voltage dips and voltage fluctuations on power systems arising from nonlinear devices connected to the a.c. supply, which becomes acute with the larger units, need special attention. Intermittent voltage fluctuation due to change of load pattern significantly affects the quality of supply when interferes with other loads. The effect of voltage variations on power systems due to fluctuating loads are to be evaluated on case-to-case basis as it depends on magnitude, frequency of occurrences, and the rapidity with which the changes take place. The frequency of operation of most popular booster synchrotrons is very low for which it is essential for the system designer to take care that it does not disturb the feeding mains more than the permissible limits of the power regulatory authority. The power supplies have to deliver the power at the required frequency with high accuracy. In this regard, implementation of standards set out by each country is required to be strictly followed. For instance, no energy consumer is allowed to produce disturbances more than 3% of the voltage variation of the mains according to IEC38/VDE 0838 of the equivalent European standard EN 6100. The load patterns get affected because the magnitude of reactive power drawn from the supply for the converter is directly proportional to d.c. load current and inversely proportional to d.c. voltage during rectifying or inverting. Adequate reactive compensators or energy storage elements or combination of both are required to be installed to reduce the active power swing to protect the sensitive loads connected to the neighboring consumers from the source power supply. The voltage fluctuation due to rectifier loads are either supplied at increased cost from the points on the networks of high short circuit level or adequate reactive compensators and static VAR compensators (SVCs) are provided at the bus bar to minimize this problem.

Grid-friendly power supply topologies with active front ends are now gaining interest. The identification of the point of common coupling (PCC) is important from the utility point of view and usually, it is either the point where power comes into the establishment or the point between the nonlinear load and other loads. IEEE standard 519-1992 relies strongly on the definition of the points of common coupling (PCC), which from the utility point of view will usually be the point where power comes into the establishment, i.e., the metering point. Actually, the quantum of harmonics generated from controlled or uncontrolled rectifiers needs to be assessed for mitigating harmonics at or below the acceptable level. In multi-pulse converters, several techniques are adopted to make the input current waveform continuous. The overall current harmonic distortion is reduced to some extent by using transformers with phase-shifted outputs. Some other frequently used techniques for reducing harmonics and ripples are the introduction of line reactors of significant magnitude at the front end, phase multiplication and application of passive and active filters. In converter design, the phase multiplication is selected after knowing the characteristic harmonics from the expression $h = np \pm 1$, where h is the harmonic number, n is an integer, and p is the pulse number of a full wave rectifier bridge. The above functional relation confirms that the lowest harmonic generated by a 12 pulse converter is 11 and of 24 pulse converter is 23rd. The cost of construction of a 12 or 24 pulse unit is higher than that of an equivalent 6 pulse unit employing parallel bridges but interestingly, the cost for harmonic filter for 12 pulse converter, is much lower than that needed for 6 pulse converter. The two bridges used in a normal way would eliminate the fifth and seventh harmonics any way by virtue of 12 pulse operation. Actually, before choosing an acceptable harmonic filter the simple question required to be asked whether the new method will give better result in comparison to known methods. For instance, when it is sufficient to eliminate the lowest harmonic and reduce the higher harmonics in multiple converters by conventional means, a method of triple frequency injection is suggested at times to replace the static filter circuits. Economic analysis comparing the advantages and disadvantages of different filters is required before accepting the suitable one. The power factor correction capability of static filter is required to be compared with the harmonic injection type where power factor correction as well as harmonic reduction is obtained. Voltage notching due to variation in overlap period (μ), and firing angle delay (α) of thyristors, are other factors during operation which can cause major disturbances in sensitive electronic equipment that rely on the zero crossing of the voltage, especially the effect is more during multiple pseudo zero crossing of the voltage waveform and needs to be addressed to rectify such problems. The use of passive filters are common, but it has its performance limitations in reducing harmonic problems while, active filters are quite effective in reducing residual ripple contents and considered for improving the performance of the system at lower frequencies. With the progress in the capacity and switching speed of semiconductor devices, such as IGBTs along with parallel diodes, the interest in the use of active filters for power conditioning and for more practical applications has largely increased. But, active filter topologies are complicated as it involves increase in cost because it requires monitoring and control of currents and voltages continuously for which effective computing is necessary. Schematic of a widely used three-phase thyristor converter bridge network with line inductor for reducing harmonics and minimum filter network for the reduction of ripple content is shown in Fig. 5.1a. The power factor of the SCR converter changes according to the variations of the firing angle α and in order to keep the power factor close to unity, calculated reactive power compensators are included in the system. The SCR is widely accepted in converter topology for high power applications, mainly above hundreds of kW, and frequently chosen for the bending magnets of storage ring. The different structures of semiconductor devices are drawn alongside only to show that alternative

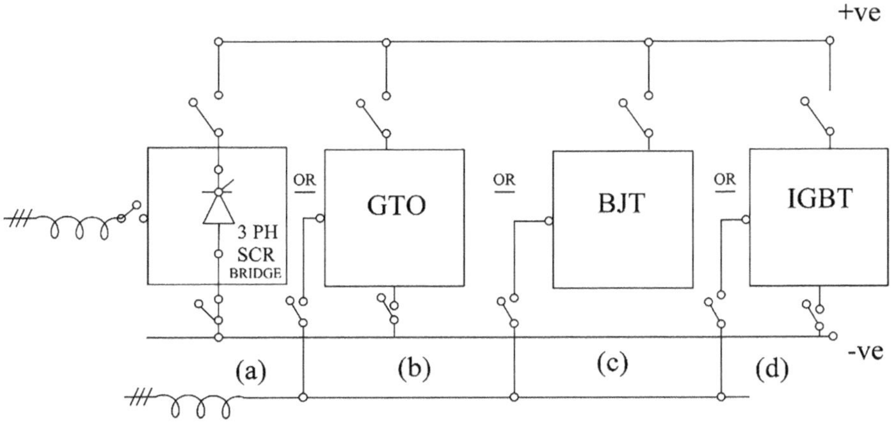

Fig. 5.1 (**a–d**) Widely used 3 ph converter network with thyristor and other devices

possibilities for static converter can also be adopted as shown in Fig. 5.1b–d. The d.c. voltage available at the output of the thyristor converter network is

$$E_0 = \frac{3\sqrt{3}E\cos\alpha}{\pi} - \frac{3I(\text{d.c.})\omega L}{\pi} \tag{5.2}$$

where

$E = Pk{\cdot}L/N$ voltage
I (d.c.) = d.c. output current
α = firing angle
$\omega = 2\pi f$
L = inductance

The topology of line commutated inverter is utilized in converting d.c. to a.c. of any desired voltage and frequency by delaying the SCR firing angle ($\alpha > 180°$) corresponding to the phase of the utility voltage. The application of thyristors in accelerators are different; the natural commutated thyristors are suitable for high power a.c–d.c. conversion and forced commutated inverter grade thyristors are utilized in d.c–d.c. choppers and d.c–a.c. inverters where turn off is done by forcing the current zero using external commutation circuits. In inverter design, the most important performance criteria are the energy conversion efficiency; therefore, switching circuits with low losses gets the preference. In modern inverters, Insulated Gate Bipolar Transistors (IGBTs) are preferred as the main power control devices, but power MOSFETs are also used especially in lower voltages in lower power ratings and in applications that require high efficiency and high switching frequencies. In recent years, IGBTs, MOSFETs, and their control and protection circuitry have made significant progress in the application of inverters. The application of MOSFET in the design of a.c–d.c. converter with high stability has been discussed (Fisher 1989). The expanded form of a commonly used topology of a three-phase inverter with IGBT as shown in Fig. 5.2 is basically an extension of the H bridge single-phase inverter with an additional part to from a three-phase circuit. Each power control device has a diode connected in anti-parallel to provide an alternative path for the load current in case the power switches are turned off. Static converter of different structures is developed according to the voltage or current sources. In a.c–d.c. or d.c–a.c. conversion scheme sometimes, it is necessary to design the circuit with several interconnected converter groups to form a converter unit. This is because although the converter groups themselves are complete converters, they are inadequate due to component ratings in delivering the design voltage or current requirements individually. This formation and application of groups has been discussed in the setup of earlier synchrotron supplies with Hg–arc rectifiers in RHEL-Nimrod, CERN, Bevatron, and other units. Sometimes, it is necessary to consider converter configurations to be multilevel with available semiconductor devices arranged as per the design demand with the objective of gaining two advantages:

1. To meet the demand of higher voltage–higher current capabilities of components
2. To reduce the harmonic content of the output waveform

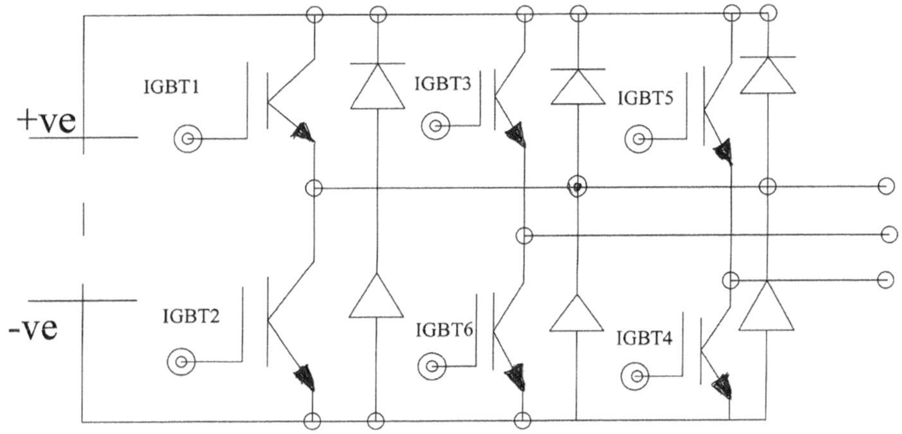

Fig. 5.2 Topology of an expanded 3 ph inverter with IGBT

The increase in the number of suitably structured switching layers is to overcome the limitation of component ratings and extract the benefits from additional layers with the added advantage of enhancing circuit redundancy. However, as with many protection systems it is essential that the design must ensure a very high degree of reliability in initial fault finding and subsequent correction even during converter malfunctioning. Actually, the design must also take care of the fact that the converter passes through the number of numerous modes of operation with an increase of current from no load to full load. The sequence and type of these modes being the functions of the coupling factor, magnitude and sign depend on whether the converter is rectifying or inverting. The use of solid-state circuitry in the multilevel application makes it possible to achieve high reliability and the fault detection, segregation, and compliance with redundant circuits are fast to resume the operation. The trend among semiconductor converter manufacturers is to build standardized converter modules arranged in series or parallel combinations using reliable rectifying elements of moderate cost and rating so that the required rating for any particular application is available off the shelf and tuned at site to achieve the required rating and stability which is most important in this case. Such multi-group converters are extensively used in industrial drives. "White Circuit," the basis of traditional resonant magnet excitation, uses two independent power supplies, a.c. and d.c. Single-phase low frequency generated by PWM inverter or by switch mode converter with the availability of high power IGBT or even single-phase a.c. supply from motor–alternator set at line frequency (RHEL) is supplied to the rigidly coupled energy storage choke of White circuit. The increasing availability of high power IGBT-driven switch mode converter technique has the advantage of switching current at higher frequencies for supplying d.c. power to the ring magnets. It is also a preferred choice for booster synchrotron which requires high voltage and high current. The converter topologies adopted for magnet excitations for mechanical and inductive energy storage are well accepted, but the capacitive energy storage and direct supply configuration to accelerator magnets were not considered earlier due to nonavailability of suitable capacitors and adequate rigidity of the supply network, respectively, for such applications. The deficiencies of electrolyte capacitors as a storage device as mentioned earlier is due to its rating limitation in high voltage applications. But, subsequent availability of high voltage capacitors as an industrial product asserted the use of capacitor to be an important choice in storing the magnet energy which is capable of sustaining a large number of charging-discharging cycles with minimum maintenance. It is now included in the evaluation process to find it economically competitive with other schemes. The application of electrolytic capacitor in switch mode system inspired confidence for adopting the scheme for the projects based on the capacitive energy storage and direct supply of power to the converters. Commercial viability of capacitor technology is expected to improve further as report of successful operation in accelerators pours in and any decision in favor of this technique needs further comparison with other available topologies. As mentioned earlier, converter topologies for magnet supply to different accelerators may vary but in principle, the design for main topologies and the methodology for stabilizing them are similar.

Control System Strategies

Control system strategies are developed on the basis of what is to be controlled and how effectively it is to be controlled. This makes the control system design for synchrotron accelerator a crucial functional component because in practice, the system requires minute monitoring of the time sequence of each activity shown earlier for one complete cycle of RHEL accelerator to maintain the synchrotron operation. Similar activity schedule are prepared for all accelerators in operation. Power supplies for magnet excitations are usually incorporating the complex scheme of converters and control system in a form that can be optimally suited and constructed for the regulation of magnet loads as per the stringent design and developed in a systematic way to follow the specification laid down for accelerator operation. A number of individual control loops such as r.f. generation, heat extraction, maintaining correct vacuum in the torus of the magnet, monitoring radiation level and power supplies to the magnets, to name a few, are important in the design for accelerator operation. Converter topologies for the two main loops: one being the most critical loop for the r.f. generation has been described before and the other loop dedicated for the accelerator magnets; booster and storage ring are given special importance in the design because of their complexities in the selection from different types of magnet excitation arrangements and their contributions in the functioning of accelerators. In White circuit, the combined functions of the magnet excitation and energy storage mechanism are tied up to form an essential part of an accelerator system. The selection of a particular converter topology becomes central in decision-making, and it is necessary to compare available alternatives in selecting required excitation current as per the demand, considering their boundaries of operations with respect to reliability, accuracy, efficiency, performance, and cost. Electrical isolation of the devices under normal operating conditions requires proper selection of components and configurations, and this is important in the design of customized control circuitry to extract the best performance. There are three kinds of power supply group for accelerator magnets classified according to their operational requirements; d.c. supplies, a.c. supplies, and pulsed supplies. d.c. power supplies are required for r.f. circuits and lattice magnets; dipole, quadrupole, sextupole, and corrector magnets in the main storage ring. The a.c. power is supplied to the booster synchrotron which uses dipole and quadrupole magnets with two objectives: first one is to make up for the a.c. power losses in the energy storage choke and magnets in White circuit, where a d.c. bias current is added to the sine waveform and the second one is to initiate the synchronization. The pulsed power is supplied during injection and for specific magnet loads.

Three well-accepted basic power regulator topologies for supplying desired current and waveform to the magnet load as per the requirements are shown in Fig. 5.3:

1. Linear Mode Power Supplies (LMPS)
   (a) Series regulator type
   (b) Parallel regulator type
2. Hard switching mode converters
   (a) Switched Mode Power Supplies (SMPS)
   (b) Pulse Width Modulated (PWM) power supplies
   (c) Switched mode converters
3. Soft switching or resonant type

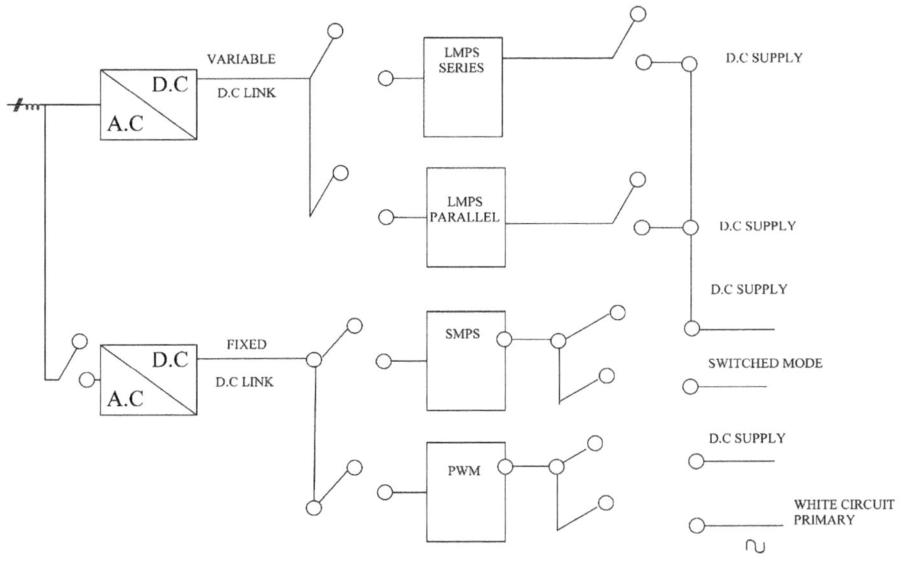

Fig. 5.3 Power regulator topologies

Switched mode converters and soft switching or resonant type converters are developed much later.

LMPS: Series Regulator Type

In Linear Mode Power Supply—series regulator type is popular and as the name implies variable voltage d.c. supply from the rectifier output is connected to regulator units in series to the main current flow and driven in linear mode. The efficiency of such a topology is low since losses are proportional to the current needed by the load and the effect of thermal variation in power transistors placed in series being prominent; its application is limited only for low power supplies.

LMPS: Parallel Regulator Type

LMPS—parallel regulators as alternative to series regulators, eases the problem to a large extent. But, these power supplies are not usually bought as off the shelf items, instead efforts are made to develop in-house as per the requirements. In-house development of power supplies of more than 100 KWs were taken up at Rutherford Laboratory during 1960s to support various magnet loads along with the bought out d.c. stabilized power supplies (Dey 1969). The possibilities of thermal mismatch of power transistors connected in parallel aided by the size of the line frequency transformer and LC filter network also pulled down the operational efficiency to certain extent due to heat dissipated in regulators despite adequate water cooling arrangements provided for the heatsinks. Further search to find suitable topologies to reduce power losses in regulators has improved the operational performance.

SMPS

The basic principle of Switch Mode Power Supply (SMPS) topology is to chop the fixed voltage d.c. link available from rectifier output to function as a chopper to produce an average output voltage constant by controlling the train of output pulses of same amplitude. The objective is to minimize switching and conduction losses by increasing the switching frequency to keep the voltage constant. The advantage of switch mode power supplies is in the choice of frequency of operation, its high efficiency due to low losses resulting from better use of the magnetic materials offer good regulation depending on the type of topology. It is highly preferred and very common for medium power applications because of small volumes. The line frequency in this case, no longer remains the synchronizing and commutating element.

PWM

Pulse Width Modulated topology, similar to SMPS, worked on fixed voltage d.c. link output. In both cases, conversion from d.c. to a.c. implies that inverter bridges operate from constant voltage d.c. link. The operation of SMPS has been explained, but the principle of PWM is different, where the d.c. link output voltage is effectively chopped up into discrete parts consisting of constant amplitude variable width blocks which produce a sinusoidal waveform when supplied to an inductance. The average value of voltage and current of the inverter is dictated by the switching frequency which needs to be high enough to produce stable and smooth current waveform, virtually a pure sine function. The popularity of PWM topology has grown largely from the demand generated by industrial sector for variable speed drive in improving the performance of a process requiring maximum reliability in comparison to other methods. In VSD (variable speed drive), the typical converter control application is hard-switched voltage source inverter producing pulse width modulated signals with sinusoidal fundamentals. The generation of sinusoidal waveform for the a.c. supply source to White circuit by PWM topology also considers voltage source inverters (VSI) and operate at very high switching frequencies with controllable power semiconductors, where current and voltages are limited due to component ratings. Converter circuits requiring high reproducibility, low current, and low voltage on the d.c. side are often designed with PWM topology where hybrid structures are considered to be a preferred solution for this application to satisfy the demand of high precision and fast response. The PWM control signal is also generated using modern microcontroller and DSP and these digital controllers are for high performance dynamic response.

Switched Mode Converters

Converter topology being the key technology for accelerator operation has brought a drastic change in the conception of magnet excitation with new found semiconductor devices. The switched mode configuration replaced the technique of mechanical or electrical changeover switches with a fast-acting interrupter, an improved version of switching configuration between supply side and magnet for particle accelerator application. The design of switched mode converters (Dupaquier and Proudlock 1990; Isch 1989) using transistors as interrupter between their off state and saturation mode emphasizes to minimize losses in mid or high power circuits. The structure of switched mode converter as shown in Fig. 5.4 is developed in several stages and essentially consists of a fixed voltage d.c. bus (converting a.c. to d.c), d.c. chopper, energy storage capacitor, switch mode controller with IGBTs (Q_1 and Q_2), and output filter network to supply power to the magnet designed as single module and shown in the simplified form in Fig. 5.5.

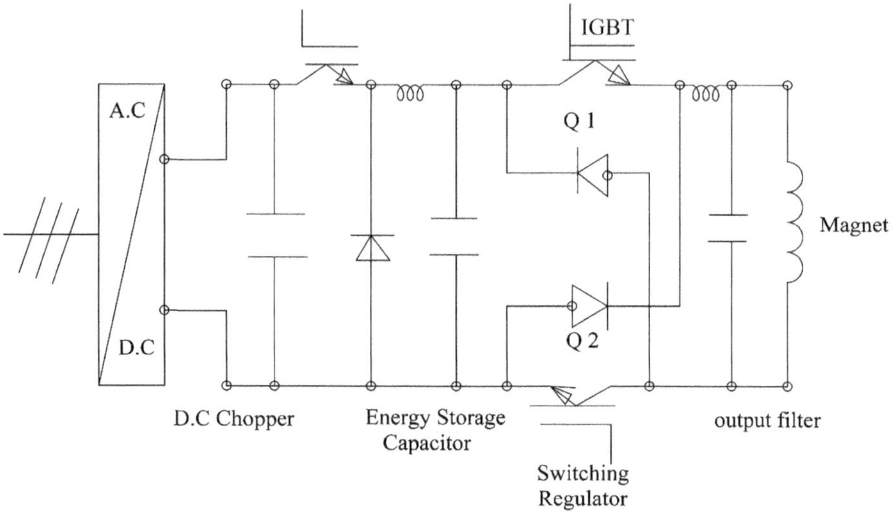

Fig. 5.4 Structure of switched mode converter–schematic

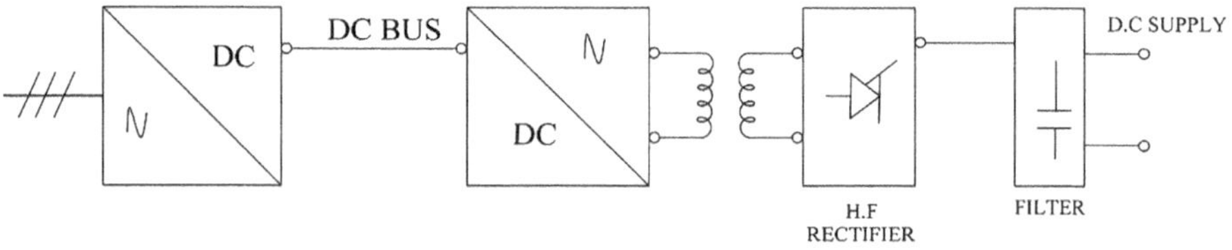

Fig. 5.5 Switched mode converter with IGBT

Depending upon the conducting state of the two IGBTs, the magnet load is either connected directly or inversely across the energy storage capacitor or remains in freewheeling mode. Such functioning is highly favorable when charging—discharging duties of suitable capacitors are found to be acceptable. In fact, this scheme with IGBT a semiconductor device in the package filled the long awaited gap in our search for a change in the pattern of magnet excitation. The reason for accepting customized switched mode converter with IGBT and electrolytic capacitor as a package greatly enhanced the popularity of this scheme and changed pattern for high power magnet excitation offering significant advantages over inductive storage.

Switched mode converter was designed and developed by PSI (Paul Scherrer Institute, Switzerland) with high voltage electrolytic capacitor bank and established this configuration to be the pioneer after successful installation in various projects. In order to achieve higher power, several of these power supply modules with IGBTs and capacitors are grouped in stacks to achieve the required power (desired range of voltages and currents) by connecting the modules and the capacitors in series or in parallel groups. Such groupings ensure that total excitation current to magnet circuit is the sum of the currents generated by each power converter unit with the possibility to nullify any small current variations developed in the circuit to supply the exact magnet power. The modular approach provides the facility of ease in maintenance by switching over to redundant module as an intermediate support to take care of the immediate fault without disturbing much of the accelerator time and continue smooth and uninterrupted operation when replacement of faulty module is taken up at later stage. The flow sheet of charged particles in the basic accelerator structure with Linac, Booster, and Storage ring connected in tandem is shown earlier where addition of a storage synchrotron draws particular attention which distinguishes the difference. The booster synchrotrons are cycling accelerators operating in the frequency range from 0.5 to few Hz. In this new structure, electrons from Linac are injected with sufficient initial energy to booster synchrotron where a specifically designed r.f. cavity raises the energy of the charged particles as they circulate inside the torus of the magnet to attend the specified level of the beam energy. Thereafter, the particles are transferred to a storage ring either

- to raise energy to higher level
- or utilize the synchrotron as storage unit without raising further energies to the particles and lock it for certain period so that the beam can be extracted at suitable time as per the demand
- or else, it may be of continuous topping up type supplying beam for the experiments depending on the facilities offered by the accelerator complex

Actually, the switched mode technology is well established for ring magnet power supplies which require constant current. But, special design configurations with most innovative architecture of modules are introduced for variable current requirements and ease of maintenance for booster power supplies. Whatever may be the type of switch be it thyristor, GTO, BJT, MOSFET, or IGBT, the excitation waveform should remain the same. In many projects where high voltage and high current both are required, the application of mixed technology such as thyristor (SCR) with switch mode active filter is adopted to improve the performance. In the design of high frequency switching mode power supplies for the high power application, increased switching losses results when the converter operates under hard switching conditions. This is because, in the switching technology, the finite duration of the switching transistors will cause peak pulse of power dissipation in the device and degrade the converter efficiency. This may also cause possible damage of transistors during turn off which adversely affects the efficiency. Expert system-based switched mode power supply design is now available which is expected to give better results. Otherwise, soft switching techniques are preferred for the converters as alternative to hard switching to reduce switching losses, improved efficiency and stability.

Soft Switching Technology

A more effective solution in reducing the switching losses is sorted out by using resonant power conversion like zero current switching (ZCS) or zero voltage switching (ZVS). The distinct difference between the two is that, in case of zero voltage switching, switches are operated when the voltage is zero but, in case of zero current switching, switches operate when the voltage and current both are zero. Recently, the technology of soft switching has become more interesting mainly for reasons of electromagnetic compatibility. The soft switching inverters have an oscillatory intermediate circuit and attempt to open and close the power switches under zero voltage or zero current conditions.

So far, the available topologies and their advantages, disadvantages, limitations with respect to operational frequency and power have been discussed. It has been emphasized at all levels that the performance of the power converter and its output regulation directly impacts the beam quality for which, accurate monitoring of magnet current is very important and it is essential to supply the current as per the accuracy and tolerance level specified in the design. Power converters are always of the current control type; therefore, the control system is primarily responsible for providing few essential services:

- for initializing the central timer
- Synchronization, tracking and monitoring of current signals to provide a robust feedback loop

Synchronizing the magnet current to dipoles, quadrupoles, and sextupoles as per the time sequence mentioned earlier is again an arduous task to the control system designer where operators are tracking and monitoring the program remotely from control station. The important function of the control circuit is to ensure stringent current accuracies in magnet systems as per the timing schedule for which a reference signal is essential and in this regard, usually, the dipole magnet is considered to serve the purpose as master controller which takes the lead as reference for controlling all other magnets. The flexibility needed in generating many different types of magnet current cycles for bending magnet regulation and control is complicated and involved different sequential voltage changes required at the converter stations. The reference is generated with digital and analog computation equipped to follow set programs as per the three element characteristics mentioned earlier as the basis of accelerator cycle:

1. During most of the rise time, the rectifier output voltage is the important parameter to be considered. Hence, during injection as well as during flat top, current is the controlled quantity.
2. During decay, the converter voltage is controlled. Accordingly, the computer will generate voltage references. The magnetic field or current will be monitored by the computer so that correction can be applied to the various references during and between pulses in order to obtain the required voltage distribution and output current waveshapes.

At the start of each new injection cycle, the trigger signal from sine generator initialize the timing sequence originated from the control station and supplied to the dipole power converter. After the initiation of trigger signal, the dipole converter starts increasing beam energy as a reference to drive the quadrupole and sextupole magnets. The power supplies of all other magnets for synchronized operation are monitored and controlled by the zero crossing detectors placed at the output of dipole magnets. To design a high precision power converter, a device capable of precisely measuring the output current is selected from Table 5.1 as per the requirement.

CTs and shunt type current transducers are very often used in industrial applications, whereas quality transducers of high accuracy and principle of zero flux DCCT (Direct current-Current transducer) are capable of tracking and giving precise measurement of the output current which produces voltage proportional to current essential for closed loop control and considered as most effective than other forms of current measurements such as shunt type transducers based on Ohm's law or Hall Effect devices) (Fernqvist 2006). For instance, apart from various other applications, DCCTs are preferably used in critical control systems in mines such as "Mine winder drives" where the control of the skips (service shaft or production shaft) as per the statutory safety guidelines of mines for men and material transportation at exact location is important. Similarly, shunt type transducers are used in process industries such as d.c. current (10 kA) control in copper refineries where the temperature of the electrolyte in a large tank house holding anode–cathode is required to be maintained at desired temperature and variation is only allowed to, ±0.5% but due to short circuit of anode–cathode placed inside the tank with a limited gap between the two, the possibilities of localized heating raises the electrolyte temperature beyond the allowable tolerance for which the production suffers. My working experience while working in mining and refineries with DCCT and shunt type transducer in control system applications in both mines and plants was highly satisfactory. DCCT's electronic circuit is considered to be the heart of the device and highly sensitive to external disturbances such as mechanical vibrations, electrical noises, and temperature variations. Although it is available in secured sealed packages, it needs to be carefully handled. DCCTs are preferred because the best devices are designed to reach high precision, a few "ppm" level of stability and repeatability as shown in the table and adopted in control applications by almost all power supplies for its reliability and quality performance for maintaining a stable linear relationship of input and output. In the functioning of White circuit, it is necessary to track, monitor, and control the two independent currents d.c. and a.c. separately and

Table 5.1 Comparison of current transducers

Type of variants	DCCT	Hall effect	CTs	Shunts
Principle	Zero flux device	Hall effect principle	Faraday's law	Ohm's law
Output	Voltage or current	Voltage or current	Voltage	Voltage
Accuracy	Best devices can reach a few ppm stability and repeatability	Best devices can reach 0.1%	Typically not better than 1%	Suitable for low currents <0 for high currents
Ratings	50 A to 20 kA	100 A to 10 kA	50 A to 20 kA	mA to several kA
Isolation	Yes	Yes	Yes	No

control the required current in a resonant magnet excitation circuit. The output proportional to a.c. peak field is converted to d.c. voltage and compared with the set point, the result of which is fed to the input stage controlling the sine generator. The output of DCCT is fed to the input controller to meet the accuracy of current demand. The resultant output regulates and controls the power supply to magnets by using PWM topology as shown in the schematic diagram 5.6. Synchronized control to maintain precise execution of the timing sequence generated from the central station as indicated in time cycle graph of Nimrod or of any other machine is vital to the functioning of all power converters and considered as the most important task for the accelerator control system designer. Analog regulation is considered adequate in accelerator control due to its choice of simple circuitry, excellent quality of components with high level of reliability and performance over a long period at reasonable cost.

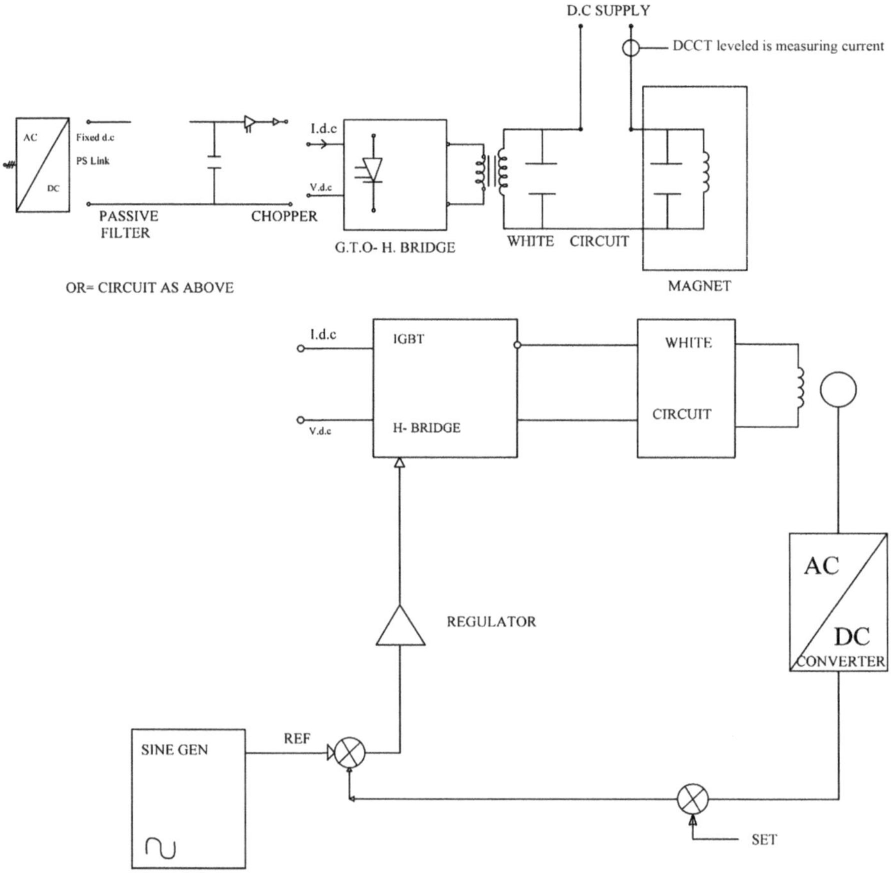

Fig. 5.6 Control schematics—power supply to magnets

The pattern of control systems has changed in multiple fields over the years but, the remarkable transformation especially, a quantum jump in control system applications from primitive methods has been discussed at length elsewhere to highlight how advancement in technology has helped in the progress. The higher demand of tolerances, accuracy in current control tied up with the beam dynamics for accelerators invite more sophisticated control against conventional type of analog control to give improved performance in many cases. The design and availability of circuits for digital regulation with latest state-of-the-art components is becoming popular (Bordy 2006) because it is proven that the digital techniques as against conventional type of analog control are expected to improve performance. The control system designer is now faced with a wide choice from available software packages for each task within the design process.

References

Bordy F (2006) Regulation theory—review and digital regulator. CAS

Bouterille JF (2015) Power converters for cycling machine. ESRF, Grenoble

Burnet JP (2016) Requirements for power converters

Dey PK (1969) The distribution and utilization of electrical power at Rutherford high energy laboratory. IEE SQJ (Lond):1–7

Dupaquier A, Proudlock P (1990) Switch mode converter

Fernqvist G (2006) High precision measurements. CAS Daresbury Laboratory, Warrington, p 299

Fisher A (1989) An a.c.–d.c. converter using Mosfets for a 2000A, 10V output with high stability. In: Proceedings of the 3rd EPE conference, Achen, vol 3

Isch HW (1989) Switch mode converters, present and future. In: Proceedings of the PAC conference, Chicago, Mar 1989

Marks N (1996) The choice of power converter systems for a 3 Gev booster synchrotron

Marks N (2004) Power converters for synchrotron accelerators. CERN Accelerator School (CAS), Warrington

Chapter 6
Miniature Model of Developing Functional Structure of Accelerator Project

Preparatory Stage

So far, power supply technology, changes in the pattern of magnet excitation and the controls for synchrotron accelerators are briefly described. But, how such accelerator projects are initiated and developed needs to be briefly explored. In this regard, it must be emphasized that a strong functional structure for planning, monitoring, and coordination with a comprehensive approach to undertake project control, construction, and operation of a large synchrotron with hundreds of different activities are required to be strictly and minutely followed for completing the task within the specified time frame. Synchrotron accelerator project is comparable to Philharmonic Orchestra where each musician touches the perfect chord and completes the task within the given time frame to create a memorable symphony at the end. Primary function of initiating a synchrotron accelerator project is very similar to building a good team of musicians capable of performing their specific tasks for the orchestra. A functional structure is necessary to serve two functions: one during the project construction stage and the other during operational stage and the two stages need to be fully coordinated to serve the desired objective of the project facilities. It is a huge task even then few important assignments are only listed for preliminary consideration before finalizing an accelerator project such as land, power, road transport, institutional and medical facilities for those countries aspiring to develop their own accelerator.

- Land requirement—Land requirements for accelerator project needs to be assessed in two phases firstly, at the project initiation stage and secondly, for future plans of expanding experimental areas.

Criteria for land selection:

- Preferably, easy accessibility by road and by train from nearby airport.
- It should preferably be near some reputed science and engineering institution.

P. K. Dey, *Power Supplies for Synchrotron Accelerators*, Springer Tracts in Electrical and Electronics Engineering,
https://doi.org/10.1007/978-981-97-7606-1_6

- Availability of high voltage grid connections from dual sources.
- The township to be developed for housing the experts and for the technical and nontechnical supporting staffs. It is expected that a large number of foreign experts including the members from joint collaborators will be visiting the site frequently for participating in research work.

The project site for proposed accelerator will not only cater to scientific experiments on nuclear physics, but it should also procure land for further expansion of facilities for new development purposes to explore the possibility of changes or addition of the program of scientific research in future on emerging technologies. It is possible that the program of scientific research may also change its directions with time as mentioned below:

- The possibility of raising energy levels in future
- To study the effect on high voltage grid for direct supply to the magnet system and test the acceptable tolerance of the grid due to voltage fluctuation and quantify additional space for providing system corrections
- Provide enhance healthcare facilities of medical utility unit for continuing research on the findings of experiments towards medical benefits
- Provide computer center for evaluation of scientific data available from experiments
- Workshop facilities for electrical, mechanical, civil, and instrumentation jobs for in-house support
- Heat transfer and cooling water arrangements
- Administrative buildings
- Training center

To reduce time, cost, and effective control of an accelerator project, the importance of a sample functional structure needs to be specifically prepared with various activities essential for monitoring and coordination between different disciplines. The application of PERT/CPM (Project Evaluation and Review Technique/Critical Path Methods) and to take advantage of project management, Expert Knowledge-Based System may also be implemented to support such functional structure for effective construction, maintenance monitoring, and operation control. PERT/CPM technique for project management is proven and well-accepted method for controlling and reducing the project cost and time, whereas expert planning system is a modern method offering various advantages including cost benefit. It is very difficult for the operator to follow the timing characteristics for a typical accelerator operating cycle sequence for which this knowledge-based system or in other words, expert system could very well be implemented for better performance. The setting of controller time sequence can be clubbed for adjusting the disturbances and system parameter variations like magnet current or r.f. current. Current control is very important for different magnets and rule-based computer algorithm can be developed that mimics an operator. A sample mini activity schedule developed as an example and shown below in Table 6.1 is to select and specify the task for each participants of the section engaged in the construction activities of an accelerator project.

Table 6.1 Sample activity schedule for project management

Sr. no.	Activities	Tasks
1.	Project activities	Project preparations—to find convenient location preferably nearby city, easy accessibility from airport, with road/rail connectivity, nearby university and reputed research centers. Easy availability of high voltage electrical supplies from dual sources, etc.
2	Estimation of project costs/ fund availabilities	Fund sharing—selection of collaborators/partners/sponsors/facility users. The reason for such fund sharing is purely due the exorbitant cost of such project which will be difficult for a single country to divert such fund from the allocated budget. The cost estimation will simply increase with the increase in energy requirement and the type of research to be continued
3	Project design	Select accelerator team to develop process flow sheet of charge particles according to the scientific facilities to be provided; prepare magnet design data; build-up engineering design team with civil, mechanical, electrical, electronics, instrumentation, control and computer engineers Engineering team will study the total requirement for physical, functional, and operational design Civil engineers to design and develop the accelerator complex and initiate constructional works Mechanical engineers will design the magnet system as per the accelerator requirements for physical and operational design. Select manufacturers, contractors, inspectors, check on material specifications with regard to fabrication, weld design for magnet. Installation of heat transfer equipment, vacuum systems, and air conditioning plants. Prepare schedules for inspections, installations, commissioning, and testing 1. Inspection (a) Fabrication and procurement (b) Statutory inspections (c) Periodical inspections (d) Any special requirements to check out on physical and operational design during service Electrical engineering team with electronics, instrumentation, control and computer engineers will study the requirements, design the power supplies of magnet systems, select manufactures, contractors for various jobs such as monitoring and control of magnet currents, r.f. system, heat extraction from various sources, monitor and control of vacuum systems, air conditioning plants, etc. Negotiate with power supply authorities Similar inspection with regard to electrical equipment at the manufacturer's site. Monitoring of safety requirements 1. Inspection (a) Check on limitations of specifications (b) There are many such factors which are not mentioned in this short list

The tender preparation stage for such a system of accelerator project is complex in nature due to the standard of precision required during installation, commissioning, testing, and the frequent inspection clause with regard to monitoring of the delivery dead line. Hence, the tender document needs to be prepared considering all the aspects prepared at the planning stage in the list which is lengthy and an expensive process. The tender documents are to be served only to the reputed parties having experience in the construction of accelerator projects so that their feedback can be well utilized. The specification problems are ironed out after regular discussions with design team and the vendors as far as possible before the tender documents are issued. Any clarifications against specification are to be made before the invitation of tender after working closely with the vendor. To oversee and monitor the progress of project construction activities, some procedures are usually followed for which a functional structure is essential as guidance. This activity list can be extended to demonstrate only how critical points of the particle accelerator need to be focused by segregating the responsibilities of each section. To complete a project in time, this type of activities with specific pointers can be expanded to any extent in a systematic way depending on the size and the facilities to be offered.

Execution Stage

As an extension to the short mini activity list presented earlier, the minimum functional organization structure as shown in Fig. 6.1 is developed to segregate the essential areas by isolating magnet physics, engineering division, and the experimental zones to stress on the need of proper connectivity and effective coordination of relevant disciplines without sacrificing the services and facilities of the accelerator.

In Fig. 6.1, the Project Director is comparable to the Conductor of the Philharmonic Orchestra. Actually, a distinct partition by segregating planning, designing, execution, and beam delivery mechanism is expected to improve coordination and timely completion of the project to keep the cost under control. At one end, the magnet physics and accelerator division concentrate on the particle injection mechanism and peripherals, draw-up specifications for the accelerator, select energy level aspects based on the specific application, set the value of initial beam energy to be injected from the LINAC, fix up initial and final magnetic field values, quantify the size of the accelerating structure with its radius, decide on magnet parameters, type of magnet orientation, specify the number of r.f. generators to be installed, the pattern of magnet excitation to be adopted and take the most vital decision in choosing the operating cycle frequency of the accelerator as it dominates the design and the project cost. The accelerator section prepares the construction guidelines and its limitations, specifies the required vacuum for the accelerator, and considers all other demands of the magnet physics before raising requisitions and forwarding them to different engineering disciplines to study and check the given specification by each section before freezing the final design. The engineering division at their end takes the decisions at the initial stage on essential parameters such as the composition and arrangement of magnet systems, the load characteristics, the level of beam intensity with minimum losses and the allowable stability limits with the highest acceptable tolerances to initiate negotiations with manufacturers and suppliers. The strategy is to guide and interact at regular intervals with the outside agencies and manufacturers of different categories, negotiate with suppliers,

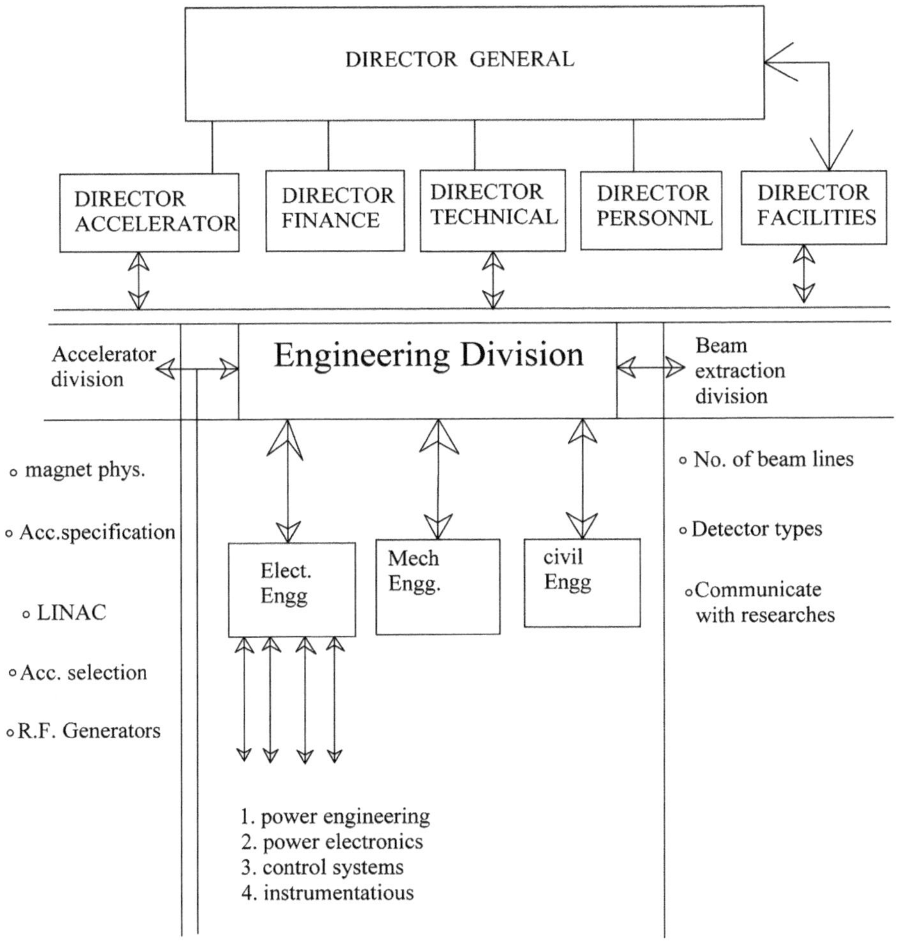

Fig. 6.1 Functional structure of an accelerator project

contractors, and clarify the specifications to make them understand the importance of stringent tolerance levels expected in the machining of magnet parts and its alignments at the time of installations, the requirements of different power supplies to the magnets, negotiations with the power supply authorities, the importance of vacuum facilities of the highest order require during operation, air conditioning, arrangements of heat extraction scheme from r.f. generators and from other sources and their effective utilization to attempt for a total collaborative efforts by all the disciplines in the construction and development. At the extreme end of the partition is the experimental area where power supply requirements for transportation of beam, the decision for providing a maximum number of beam line magnets with beam extraction facilities such as plunging mechanism and selection on the type of detectors for different scientific program schedules for the user community by the beam handling units are taken up to ensure the strategy of beam delivery mechanism to the satisfaction of accelerator operation and evaluate its economic viabilities. This detection part is likely to be separate for each project and depends on the facilities to be offered. The segregation is essential for a large project where the contribution of all the engineering disciplines, civil, mechanical, electrical, instrumentation, and controls are closely tied up to streamline and strengthen the inter dependencies between departments to ease the design, development, construction, installation, commissioning, monitoring, modifications, and successful operation. A well-defined functional structure is all the more necessary and useful for the countries like to set up jointly, the most ambitious project of an advanced particle accelerator of their own with the restricted fund. Recent advances in the fields of medicine, chemistry, electronics, nano-technologies, and other important industrial applications have prompted the design and construction of many third generation synchrotron radiation facilities at the intermediate energy ranges of 1–3.5 GeV. The aim to develop such facility on priority is to extract abundant benefits generated from such machines, for example, impact on industrial applications cover a broad scope such as ion implantation in the semiconductor industry, modification of surface properties of materials, and the basic components for manufacturing advanced systems with high density integrated electronic circuits. The support of such collaboration, joint ventures, and participation is necessary for the researchers to continue their studies even with the participation of hired facilities to fill a huge gap that exists at the current state in research infrastructure. In the engineering division, design experience and the knowledge gathered from earlier synchrotron machines are shared by all the disciplines for guidance and correcting the future course of action to take advantages of modernizing or upgrading each engineering section including r.f. generators, converter configurations, power supplies for magnet excitations, and in other scientific and technological areas for improving performance. The advancement of particle accelerators with different energy levels at affordable cost is the result of the evolution of power electronics. In this regard, the contribution of converter topology in handling the most challenging task of regulating the beam dynamics share equal credit in such development which continued simultaneously with the changes in demand of magnet physics. The systematic development of different converter structures with their limitations of current regulations and closed loop control systems are designed to suit the demand of the magnets. Once the functional structure is set right for the accelerator project, the requisitions are prepared by the magnet physics and other sections to pass the information to engineering division comprising broadly of civil, mechanical, and electrical engineering. For civil constructions, to look into the safety aspects and other hazards are specified; for mechanical, preparation for design, fabrication, installations, and alignments of magnets conforming to the accelerator geometry and for electrical; requirement of power supplies with the stringent accuracy demands of the accelerator magnets during operation. Only a part of the huge engineering challenges on the electrical engineering requirements has been briefly shared here with emphasis on addressing the stipulations imposed by the power supply regulatory authorities, converter topologies, instrumentation, and control aspects in the operation of accelerators. The demand requisitions received from the magnet physics division are critically examined by the electrical engineering division before selecting the type of semiconductor devices, converter topologies to be adopted for power supplies, and the choice of control systems to offer the precision suitable for delivering the required result. The magnets in synchrotron accelerators, cover a wide range of field distributions in booster and storage ring to bend, steer, and store the charged particles for which the specification of loads are required to be accurate. The power supplies are developed as per the type of magnet excitation and other specifications selected at the initial stage, such as the pulse power or cyclic power, ramp rates, the magnet cycle, required beam intensity, stability, minimum and maximum current with specified accuracies. The principal loads for generating necessary magnetic flux densities depend on the requirements of currents for dipoles, quadrupoles, sextupoles, and other loads. The continuity of field current and its high stability with smooth current waveform is required in White circuit resonant magnet excitation to control the beam (White et al. 1956; Marks 2004; Burnet 2014). The beam intensity is the most crucial factor at the output; therefore, the total design is focused on the number of beam lines and the extraction mechanisms to provide the beam of required strength to the satisfaction of the users for their set experiments. In case of capacitive energy storage, the value of the available capacitors is to be selected as energy storage element capable of withstanding the transfer of the magnet energy at the switching speed demanded by the circuit. In case of direct supply, the focus is given on the power supply arrangements to negotiate with the power regulatory authorities by the engineering unit and evaluate the requirements of suitable inductors and reactive compensators to be installed at the bus bars for mitigating harmonics and effective functioning. All the constraints for powering the magnets make the selection of components and power converter topology most important. The design of magnets and requirements of power converters usually proceed in parallel, optimizations and compromises will only ensure technical and reliable operation. The demand requisition for the beam handling unit requires similar evaluation for beam line magnets and detector facilities for assisting the experimenters. Each accelerator is different even though their energy level and operational frequency is same; this is because the user facilities and the detectors offered at the delivery end depend on the research applications.

Cost Control, Effective Monitoring, and Training

The setup of functional structure is required to be flexible so that it can extend its scope and enlarge in-house capabilities to look into the possibilities of taking advantage of upgraded devices, develop new topologies, provide operational support and draw maintenance program with the changes of demand. To reduce time, cost, and effective control of an accelerator project, the importance of a sample functional structure is specifically prepared with various activities essential for monitoring and coordination between different disciplines. As mentioned earlier, it is difficult to divert huge amount of fund for nuclear research for any individual country in developing an accelerator project but, during interim period, i.e., prior to the availability of fund these countries could engage in in-house development programs of certain specialized items required for the proposed projects, in other words, try to engage in shadow practice on different activities which are vital for the skill development for accelerator operation and maintenance. The exorbitant amount of fund to encourage scientific research could be arranged through joint collaboration of the number of countries as it has been done in 1954, when several European countries joined hands to share the cost and form a structural body in developing the world's most prestigious research center CERN (Switzerland) at Geneva. The participation of scientists and engineers at various conferences from CERN establish that the very existence of CERN is not only to carry forward scientific research of its own but encourage collaborative research. Its multi-functional activities include spreading and sharing the knowledge gained directly or indirectly on modern advancements in technology through in house or outside seminars and conferences jointly and supported by reputed professional and technological institutions and global experts on varied subjects relevant to accelerator and take suggestions how to impart specialized training for skill developments. Postgraduate Education Centre, Harwell, an establishment of Atomic Energy Research Establishment (AERE), a next door campus to Rutherford High Energy Laboratory, also set up a similar center where technical conferences, seminars, and skill development programs to facilitate unperturbed accelerator operations are conducted as a part of the continuous teaching process. Few major areas are identified below for skill development exercises which will help to generate a pool of engineers capable of proper handling of the final project and efficient particle accelerator operation from electrical engineering discipline.

- Power source—which will encourage the study on the effect of direct supply connections of power sources from grid and try out emerging methods of supplying different patterns of power supply for magnetizing current. Find methods of reducing voltage fluctuations further at the grid and improved harmonic mitigation techniques.
- Develop maintenance training program on equipment for electricals—high voltage transformers, switchgears, and bushings to restrict faults during operation of accelerator. The maintenance procedures for H.V. transformers, switchgears, and bushings as per the manufacturer's guidelines and routine follow-up of proper schedule. Maintenance schedule has to be established and practiced, particularly to keep the bushings clean to get the best performance from such high voltage equipment during accelerator operation and minimize losses in experimental timings by avoiding breakdowns and power system faults.
- Development of capacitor technology suitable for Var compensation at the input sub-station bus bar, in switched mode operation and efficient stacking of capacitor banks in resonant circuit demands further improvements. We have come a long way and met the challenges with the advancement of technology in the development of capacitors to make it more compatible by minimizing the deficiencies for application in accelerator operation. Although new developments in

capacitor technology have encouraged us to use it in switched mode converter efficiently still, there exists few more scope for upgradations such as for the capacitors being used in CERN's latest design for accelerator supplies. Hence, work related to improve capacitor technology further can be taken up with the manufacturers.

- Develop superconductors for increasing the current density in magnet windings and r.f. generation. This will be another way to change the pattern of magnet excitation. It was discovered at the beginning of twentieth century that, many substances undergo a transition to a state of zero resistance at a very low temperature. As the research progressed, it became obvious that a class of superconducting materials can be realized which should be capable of sustaining high current densities in high magnetic fields. In principle, high magnetic field in the range of 20–200 kG can be produced without difficulty by passing a current through suitably shaped copper coil. But, as the magnitude and the volume of the field increases, the power requirements rapidly becomes high, and this creates wide spread application possibilities in many branches of science and technology. For instance, significant work was carried out on MHD (Magneto Hydro Dynamics) at Culham Laboratory to realize one such application. My first visit to Culham Laboratory was as a participant with the team while attending a specialized course on "High voltage engineering" organized by the Postgraduate Education Centre, Harwell. In mid 1960s, Culham Laboratory was a small campus in comparison to RHEL or AERE (Harwell) and being very near to Rush Common Hostel (a unit of AERE, RHEL, and Culham for accommodating nearly 250 scientists and engineers) in Abingdon, a small village midway between Oxford and Didcot, we used to visit there from our hostel to play tennis at weekends. The research and development work on magneto hydro dynamics always intrigued us because we were totally ignorant about this subject. It was told to us that MHD has relevant applications in applied sciences, "if a conducting fluid moves in a magnetic field, electric fields are induced in the fluid and electric current flows. The magnetic field exerts force on these currents which can modify the flow, at the same time currents may alter the magnetic field. The study of these complicated interactions is the science of MHD which combines the electromagnetic field equations of Maxwell with the equations of fluid dynamics." In some cases, the fluid media may be conducting liquid metal or highly ionized gas or plasma. Extensive work in solving complicated equations and complex problems of varied nature relevant to MHD mechanism and verify them in test beds was the challenge to the scientists and engineers. Over the years, the infrastructure of Culham has grown and it has now become prominent place in the map as the center of excellence in fusion research in the UK. Its significant research contribution in the applications on superconductors is attracting experts from all over the world. My last visit to "Culham Centre for Fusion Energy" was in 2016. The purpose of my visit was partly academic interest to see the progress and technological achievements of such research programs on material science and fusion research. In this front, it will be pertinent to mention that Rutherford Laboratory also initiated research on superconducting materials and its applications in high energy physics from early 1960. Other nuclear physics research organizations in Europe also concentrated in developing the new technology of superconductors and the result of which is the development of radio-frequency accelerator technology based on superconducting cavities made of niobium which are cooled at −271 °C. Still there are enough scope on superconducting magnet design and construction, which could be taken up for specific requirements.

- Power control devices—remarkable achievements are made with IGBT but still work on the selection of VSI and CSI can be taken up. For instance, dual PWM VSI fed rectifier inverter for high power bi-directional power flow and sinusoidal current can be tried for improving converter performance.
- Power distribution and utilization in magnets require developments of effective monitoring of currents in circuits and follow the timing sequences.
- Undertake design of current sensors and various detector facilities.
- Undertake implementation of Expert Planning system along with microprocessor control.

  The initial preparation was an important part prior to starting an accelerator project to reduce the problems as much as possible during the execution period. Subsequent to such theoretical training, emphasis is to be placed on the designs and development.
- Develop plunging mechanism system to divert the beam to the required site of testing bays.
- Design different power supplies for in house requirement for beam line magnet power supplies.

It will be quite appropriate to mention here that, with the modernization and changes in design, it is natural for the engineers to gear up to meet the technological challenges. To support and enhance the strength of engineering skill, Postgraduate Education Centre, Harwell (Oxfordshire) conducted one such short course on "Nuclear pulse counting technique" sometime in (1967), which was specially designed for their graduates for skill development. It was a mixed package, theoretical course to introduce the properties and characteristics of diffusion and field-effect transistors, linear and nonlinear pulse shaping circuits, linear amplifiers, multichannel pulse data analysis, and other requirements related to analog and digital circuit design followed by some project work. This sort of technical training course to upgrade the knowledge and familiarize with the latest and relevant technology on various subjects are imparted to the engineers as a part of the AERE's continuous program for the skill development of supporting staffs. Subsequent to such theoretical training, emphasis was placed on the design and development of some nucleonic instruments as a part of the project work. At that time, it was necessary for engineers to get the clarification, explanation and functioning of circuit elements, selection of components and their operational limitations from the experts, and upgrade their capabilities to get back rich dividends from their future assignments. This is particularly mentioned here to highlight operational procedure and the plan of action prepared by such large reputed research and development organization for their engineers to get the best performance from them in the near future. The accelerator development and its operation is multi-functional. On one side, the engineers are to be trained to handle with high voltage engineering equipment connected after PCC (Point of Common Coupling); on the other side, they have to design and check various control loops and test their stability, adjustment of timing signals as per the functional requirements. I can remember one such thrilling experience during initial skill development program conducted by RHEL in Postgraduate Education Centre, Harwell, in developing multi-channel pulse height analyzer after the course. The technique of pulse height measurement in nuclear particle counting is quite old and its purpose is to record the quantum of charge deposited at the detector end which is proportional to the distribution of radiation energy. In nuclear spectroscopy, the pulse height analyzer is the instrument used for recording the quantum of charge collected by the detector which has proportional relation to radiation energy and its distribution pattern can be viewed and analyzed by the experimenter. Actually, it sorts out the amplitude of pulses resembling to different energy levels into separate counting channel by height and store them in a memory unit. The pulses are counted according to their heights and each count is added to the channel assigned for that particular height. Effectively, it sorts out the amplitude of pulses resembling to different energy levels and store them in a memory unit. This process would involve using a very narrow voltage window, which is distinctly the difference between upper voltage level and lower voltage level of the pulse height setup by the experimenter. The distribution of energies under observation is obtained by changing the setting of the window to higher and lower values successively. Pulse height analyzer method is developed primarily for this application and other uses such as measurement of pulse–time distribution, and multi-channel scaling. Actually, the earlier method of laboriously recording spectrum of gamma-ray energies was with single channel pulse height analyzer, with pulse shaper, discriminator, linear amplifier, a counter, and a timer. A single channel analyzer records those signals which fall on the channel pre-selected with proper biasing to accept a definite portion of the energy spectrum while discarding all other signals whereas, a multi-channel analyzer is capable of discriminating pulses of different heights resembling to energy levels and store them in a memory unit for further processing. A single channel analyzer has its own advantages and used satisfactorily in the number of experiments but on economic grounds the use of multi-channel pulse height analyzer proved to be beneficial. Only after the concept of pulse height technique was fully clear the basic design was undertaken and accordingly the building blocks were drawn to depict the expected flow pattern of the design and shape of the pulse suitable for pulse counting system. Practical design aspects of approaching the goal for developing the nucleonic equipment were drawn up where stress was given on the precision, reliability of design, accuracy of the instrument, fabrication method to be adopted, and testing of each circuit as per the plan.

Few essential elements in PHA are as follows:

1. Power supply unit.
2. An input signal handling unit, which includes a discriminator, a pulse shaper, a pre-amplifier, and a linear amplifier to transform the detector signal to suitable amplitude for pulse height analysis.

3. Digital devices, i.e., digital to analog and analog to digital converter to convert the analog signal received from linear amplifier and associate each input signal to a specific amplitude channel.
4. A memory unit to store the information and subsequent processing unit which is capable of keeping track of the signals as per the amplitude channel directed by the ADC.
5. Output devices preferably both analog and digital types to assist the experimenter with the necessary information stored in the memory.

With the change of technology and the availability of each functional unit embedded in LSI chips, it is now possible to get the desired pulse height analyzer in the miniature form suitable for its mobility. With the advancement of semiconductor devices, manufacturing of very high resolution Pulse Height Analyzer, adopting large channel capacity ADC, DAC, and memory unit are now readily available. But, the excitement to develop new equipment at that time was a separate feeling. Such workshops on specialized activities can be taken up in different areas to utilize the theoretical knowledge and strengthen the work force to meet the challenges in practical situations. Similarly, new exercises can be taken up to introduce upgraded technology in developing small control loops and try out implementing expert system and test the validation in accelerator operation. Also, scientists and technologists well trained in home ground on familiar topics can jointly work with next generation in the development of energy-efficient particle accelerator with latest state of the art in technology to meet the demand of the future requirements.

References

Burnet JP (2014) Requirements of power converters

Marks N (2004) Power converters for synchrotron accelerators. CERN Accelerator School (CAS), Warrington

White MG et al (1956) A 3 BeV high intensity proton synchrotron. In: The Princeton—Pennsylvania Accelerator, CERN symposium, p 525

Chapter 7
Review of Power Supplies to Synchrotron Accelerators After 1960s

Introduction

An attempt to compare the power supply configurations and their components with some useful but limited information for operational particle accelerators of 1950s has been briefly described in Chap. 3. Such comparison was possible simply because the selected accelerators basically shared the similar design philosophy of power supply for magnet excitation. With the progress of time and advancement in science and technology, those accelerators were declared obsolete long back and replaced with the new forms of design. But, it left behind only heap of past experiences gained from outstanding research work in all fronts which ultimately remained as a guide book for the engineers and technologists to refer at the time of crisis in handling difficult problems. The strength from old expertise guided to steer the progress in the development of power supplies for magnet excitation in the right direction till date, helped to extract abundant benefits for the new schemes of accelerator projects developed at later stages. A long-awaited search for alternative schemes ultimately fulfilled the cherished dreams of improving accelerator performances and reducing cost of operation with the development of new technologies which particularly concentrated on the following areas:

1. Changes in the power supply patterns for magnet excitations
2. Changes in the technology of power conversion methods and switching devices
3. Changes in the concept of control and their applications

Based on the proven technology support, a series of accelerators adopted the alternative form in the pattern of magnet excitation structures which made outstanding achievement by making them functional within a very short period.

P. K. Dey, *Power Supplies for Synchrotron Accelerators*, Springer Tracts in Electrical and Electronics Engineering,
https://doi.org/10.1007/978-981-97-7606-1_7

Comparison

Attempt has been made to review the major accelerator development project initiated with the vast operational experience only after 1960s, considering the reliability, modifications, and the changes in technologies of the accelerator design to suit the project requirements with diverse demand of specifications avoiding misplaced comparison as far as practicable.

In order to do so, the comparison of accelerators is segregated and compared in few groups as follows:

1. Comparison of inductive energy storage
2. Comparison of inductive energy and capacitive energy storage
3. Comparison of capacitive energy storage
4. Direct supply to converters with switched mode converter and IGBT as change-over switch

Only few particle accelerators have been selected in this review from large numbers available across the world in last 50 years to share some variations and point out some distinct differences in methodology adopted in these operating synchrotrons. The smooth transition from one form of magnet excitation to another aided by advance and upgraded technology developed on multiple fields has helped and contributed enormously in accelerator operation. It all started with 3 BeV Princeton-Pennsylvania particle accelerator which stands as the testimony of the pioneering work of Prof. White and the first one of such series of new found technology to accept the changing pattern of magnet excitation with inductive energy storage for synchrotrons developed after 1960s. Daresbury of the UK was not far behind and followed this changed innovation immediately by developing 4 GeV accelerator by adopting similar technology. CERN initiated different technologies to be implemented with direct supply from grid and recorded its consequences on the supply for magnet excitation. The change of inductive energy storage and the resonant magnet technology continued for a long period with little variation in their energy levels. Apart from various differences in the selection of components, only some economic comparisons that were vital in the choice of inductive energy storage scheme for the prominent accelerators of the time are given below:

Some Economic Comparison of Accelerators with Inductive Energy Storage Scheme

Economic considerations being vital in the accelerator design, the cost played a dominant role in the selection of magnetic circuit, magnetic material, voltage sources, selection of operating frequency, capacitor selection, and other factors. Some comparisons of the series and parallel resonant circuits are made earlier but, with the development of different forms of converter topology supplying d.c. power, series resonant excitation proved to be an economic choice. The cost of an accelerator project is very high, and this is further enhanced with the choice of magnet power supply for which the selection of the type of resonant circuit components are to be considered in terms of cost. For instance, in the design of distributed resonant magnet system it is important to devote reasonable time on the evaluation, selection of each equipment, component, factor responsible for reducing the project operation and maintenance cost without sacrificing the performance. The high pulse rate practically demands that the magnetic gap be drastically reduced in volume from that used in other synchrotrons; otherwise, the magnet power problem would be excessively large. Again, the choice of operating frequency which is influenced by a number of factors some of which are technical and dependent on the selection of resonant components and some of which concerned with beam dynamics are to be given due consideration in the design. The choice of the inductive storage technology put stress mainly on the selection of correct component values of resonant circuits for developing the required operating frequency, considered to be a deciding factor for magnet excitation and defining mark for accelerator facility. Hence, before determining the most desirable operating frequency for the synchrotron, comparative studies with different frequencies are essential to find out the system losses, cost of resonant system component parameters, r.f. accelerating system cost for the required protons per pulse and the energy of the accelerator which ultimately will help in projecting the cost. One such dominant factor along with the choice of operating frequency is the magnetic material. The operating frequency of accelerator contributes to a large extent on the cost of silicon steel laminations for transformers (inductive choke energy storage unit) in White circuit. How the size selection of silicon steel laminations in transformer design played a vital role with the choice of frequency for different accelerator projects and its impact on the cost in the economic evaluation is shown in Table 7.1 as an example. These comparisons are limited to 3 GeV SSRL (10 Hz), 3 BeV Princeton (19 Hz), and 800 MeV RAL (50 Hz) and explained the share of cost in the choice of accelerator frequency.

For instance, Princeton machine reduced the cost to a large extent by selecting "0.025 in." steel laminations for the transformer of 3 BeV, 19 Hz, in comparison to "0.060 in." thick silicon steel laminations, selected for the transformer of 3 GeV, 10 Hz for injector of booster synchrotron of SSRL (Stanford Synchrotron Radiation Lab.) resonant magnet power supply. On the other side, 800 MeV, 50 Hz, RAL,

Table 7.1 Selection of silicon steel lamination with frequency

Project	Frequency	Silicon steel lamination thickness (in.)
3 GeV SSRL (Stanford Synchrotron Radiation)	10 Hz	0.060
3 BeV Princeton (Pennsylvania Synchrotron)	19 Hz	0.025
800 MeV RAL (Rutherford Appleton Laboratory)	50 Hz Power supply frequency	0.0185

utilized the inherent advantage of silicon steel laminations available as in-built in NINA choke designed to operate at power frequency. The advantage of selecting such choke lied in the design, the technology of construction, and the short delivery mechanism of large magnet core at power supply frequency. In case of NINA, it used "0.0185 in." thick laminated steel sheets of high silicon content for the transformer which brought the lamination cost of RAL significantly on the lower side. Actually, such drastic reduction in the cost of RAL basically, on the advantage of such laminated steel sheets existed and readily available as off the shelf item for power supply duties. The mechanical properties for the steel are also required to satisfy the need for precision punching of laminations to the required tolerance and low coercivity of steel is preferred for its important magnetic properties. The choice of suitable insulation for the coils in a high radiation environment is also an important factor with regard to cost and effectiveness for which the glass fiber tape is given the preference as a suitable choice in comparison to other available tapes. In addition, r.f. stations in fast-cycling accelerators require more precious longitudinal spaces in rings which increase losses to make it costly; therefore, the selection should not only consider precise tracking of r.f. power and heat removal facilities, but minimization of the cost using superconductive material in r.f. acceleration cavity is preferred. Utilization of such modern technology for r.f. needs to be considered for cost reduction. The equipment in any large accelerator facility gradually becomes obsolete after working for several years and discarded to make way for upgradation but readily available for use at some other setups at a discounted price. Since all these costly equipment were designed, fabricated, tested, and installed with high degree of precision, it is customary for the advanced projects to take a stock, list the inventories and utilize these equipment with minor modifications as far as practicable at some other places. The revamped equipment is often accommodated in the vacant spaces created by the earlier machines wherever suitable and allows the exploitation of the available equipment to lower down the project cost. The working difficulty with such equipment as NINA choke from Daresbury, replaced in the vacant space of Nimrod mentioned earlier is that in the event of a failure of any winding, the choke is required to be dismantled to replace the damaged parts. But this replacement process is time consuming due to nonavailability of spares at times which ultimately affects the research program. The above accelerators set examples and encouraged other facilities to accept White circuit principle after being convinced of the advantages offered by the new technology for magnet excitation which continued till the end of the century before it is being replaced by next viable and proven alternative.

Comparison Between Inductive and Capacitive Storage Schemes

After the initial euphoria of selecting inductive energy storage as an alternative to complex energy storage with rotational equipment for magnet excitation is over, it slowly established its popularity and superiority. This configuration dominated for a substantial period in absence of any other suitable choice and thereafter ceased its preference only, when the alternative schemes like capacitive energy storage backed up by the new found switching technology stood as a challenge which continued to be well accepted by the projects and placed ahead in the evaluation process. The capacitive energy storage specifically designed to take advantage of switched mode converter topology emerged to offer following advantages over the inductive storage and improved performance than resonant circuit arrangements.

(a) Does not require costly energy storage choke. Possibilities of drifting values of resonant frequency due to change of LC components do not arise.
(b) In medium-sized synchrotrons, it can switch on and off high conducting currents in the cycling frequency of 1–5 Hz.
(c) Provides flexibility to control the magnet waveform.
(d) The design of high frequency chopper circuit in switch mode converter, controls both the directions and rate of flow of energy between capacitor and magnet.

Review of Capacitive Energy Storage Facilities

Few vital advantages offered by the capacitive energy storage in comparison to inductive energy storage are only listed here, but how this scheme became so popular, prominent, and occupied the central stage later is quite an interesting story. In fact, while the rigorous search for alternatives for magnet excitation was continuing, an exciting physical phenomenon came to light during accelerator operation. It was observed that intense and coherent light beam, identified to be electromagnetic waves moving at relativistic velocities emitted by the charged particles from a point like source described as synchrotron radiation brought a drastic change in the conception, development, and operation of future accelerators to extract greater benefits. The beam could be bent by magnetic fields from multiple magnets, dipole, quadrupoles, and sextupoles. Synchrotron radiation thus emerged as a subject of importance and initiated further scientific investigations. The research on cutting-edge materials, availability of sophisticated technology supported by superior components has contributed in advancing the new found subject of synchrotron radiation to different heights. With the increasing facilities, the demand for accelerators of moderate energy, high intensity, and high quality beam increased to a large extent. The popularity of synchrotron radiation facilities rolled out to extract the enormous benefits expected from its contributions with other advantages resulted in expanding the number of projects with synchrotron light sources developed all over the world. The prime reason was the advantages included in the design of capacitive energy storage scheme with switched mode converter technology for which the number of accelerator projects accepted the technology promptly. The extent of popularity flared to the user community from the very beginning when the efficacy of this project was properly assessed in comparison to other available methods. It will be quite interesting to know that the popularity of this scheme reached so high that it even called back for re-evaluation of the type of magnet excitation to be adopted when the selection of schemes were at its final stage in some accelerator project. The case here is referred to 3 GeV, 5 Hz Diamond accelerator (UK), where two schemes direct supply from grid to magnets and traditional "White Circuit" were initially zeroed during evaluation from which any one scheme of magnet excitation was to be selected for powering dipole in booster circuit (Marks 1996; Dobbling 2006). Suddenly, the earlier decision was stalled before finalization of the scheme of magnet supply as the success of the operation of SLS (Switzerland) started pouring in. Diamond (UK) decided to consider the evaluation process afresh with the inclusion of new found technology designed and developed by PSI (Paul Scherrer Institute, Switzerland) and implemented at SLS with the other schemes as well. Such decision was taken because the PSI design scheme was the first bright successful synchrotron radiation facility and considered as the pioneer for synchrotron radiation applications at that time where switched mode converter power was adopted for magnet excitation. This technology was found to be attractive for Diamond accelerator (UK) in comparison to other two methods evaluated earlier and ultimately the selection of "switched mode converter" for magnet excitation of Diamond was finalized. Later on, CLS (Canada), Soleil (France), and other synchrotron projects also followed SLS technology and promptly adopted similar scheme for booster power supply. It may be a guess that there might have been two possible reasons for sudden change of decision to discard the other schemes and give priorities to accept SLS technology in final evaluations of Diamond; one reason for such decision might be exorbitant cost of installing the Var generating circuit and the storage capacitor required at the supply bus bars anticipating the subsequent effect on grid in case of direct supply, other possible reason may be the cost and operational disadvantages in the choice of operating frequency for inductive storage. Actually, such critical analysis of different schemes for magnet power supply and change of decision before the commencement of the project is quite justified and normal before huge investment is made for any project. Because at the end, the long-term performance improvement and the future power cost helps and plays an important role in taking such vital decisions. The 5 Hz Diamond booster is a full energy injector designed to accelerate an electron beam from 100 MeV to 3 GeV for the storage ring. The comparison of basic parameters for SLS (Swiss Light Source), CLS (Canadian light Source), and Diamond (UK) accelerators to distinguish their design and primary features shown in Table 7.2 reveal that the frequency of booster for Diamond being higher than the other two, both current and voltage requirements for Diamond is also higher. All such factors play an important role in the selection of storage capacitor for switched mode converter.

In accelerator application, it is now possible to procure power converters from external sources for dipole and quadrupole magnets as standard products which are now commercially available in the modular form, but very often these modules are required to be tuned, tested, and customized at site to satisfy the stringent demand of precision, accuracies, ripple content, and other factors as per the requirement of circuit design for specific accelerator application.

Table 7.2 Major differences of electrical parameters of the bending magnets with increasing order of frequency

Facilities	2.9 GeV CLS (Canada)	2.75 GeV SLS (Switzerland)	3 GeV Diamond (UK)	Units
Magnet current	1092	950	1271	Amps
Magnet voltage	200	1000	1800	Volts
Frequency	1	3	5	Hz
Inductance/Ckt.	72	80	9.1	mH
No. of circuits	4	1	1	
Maximum stored energy	42	36.1	73.5	kJ

It has been discussed earlier how CERN initiated various schemes and opened up possibilities of direct supply for magnet excitation by modifying the input circuits and judicious application of capacitors. The assessment of CERN's latest schemes still need to be time tested in comparison to performances and costs. The digital system of control has also helped to concentrate and transfer capacitive energy technology in improving performances to different level.

Adopted Methodology of Power Converters for Different Accelerator Magnets

Adopted methodology of power converters for different accelerator magnets may vary depending on the application in achieving the required accuracy demand. For instance, particle accelerators and variable speed drives are part of Power Electronics but different branches of applications, equally concern to improve the performance and efficiency of each system. Both share the common platform of semiconductor technology and converter topology, share the same interaction with mains related problems imposed by the power supply authorities, share the same type of transducer support and to some extent share the closed loop control strategies with digital and analog interfaces as per the design demand. Power converters for particle accelerators are often required to exercise constraints and satisfy a performance that is unusual in industrial applications or even in some other design context. Specific emphasis in the design of converters for particle accelerators is to control the bending and positioning of the beam around the accelerator lattice for which strong uniformity of the field amplitude and waveform between all dipoles is essential. The selection of converter topologies for powering accelerator magnets distinguishes it from variable speed drive or regenerative power circuit applications because it requires to follow few basic constraints attached to the regulator specifications and imposed conditions as listed below:

1. Dipole, Quadrupole, Sextupole power supplies require tight characteristics with respect to precision class (current resolution, residual ripple, and for short- and long-term stability).

2. A relatively high static and dynamic performance from the power converters.
3. Minimum and maximum current is required to be accurate.
4. Current stability, its reproducibility, its accuracy and its dynamics, di/dt need to be strictly maintained because any current variation directly impacts upon the beam quality, beam size, and even beam loss.
5. Beam stability, beam size, beam quality, and current variations are required to be maintained in the "ppm" ranges only.
6. The ramping type and ramping rates for linear or sinusoidal supply are selected at the initial stage and maintained during operation.

Major magnet supply stability and ripple requirements of typical accelerators shown in Table 7.3 will explain the rigid criterion that is required for the dipole and quadrupole supplies of different booster synchrotron projects.

Still, some differences exist in the structure of converter design which distinguishes the segregation. Such subtle difference between the long-term stability and the ripple factor requirement criteria designed for converter types with particle accelerators and variable speed drive applications or any other applications always exist due to the nature of their applications and the type of demand in each case.

The selection of the type of control strategy to be employed for the voltage and current supply to the magnet depends on the method of magnet excitation to be adopted, acceptable limits of power losses, maintaining required accuracies demanded for magnet current variations and on cost burden. The application of ramping excitation is limited in pulsed power because of lower repetition rates while in cyclic accelerators, ramping are fixed with the choice of frequency of operation. Fast ramping machines have short ramping times and the voltage during ramping is significantly higher than during steady state because of the term $L\frac{di}{dt}$ associated with the voltage in inductive load. "White Circuit," being the traditional circuit to power a fast-cycling synchrotron uses inductance-capacitance duo as the energy storage unit which has both advantages and disadvantages. It has been

Table 7.3 Major magnet supply stability and ripple requirements of typical accelerators

Project	Type of synchrotron	Type of magnet	I_{out} (A)	V_{out} (V)	Long-term stability (ppm)	Ripple (ppm)
Bessy-II (Germany)	Booster 1.9 GeV 12 Hz	Dipole	1375	120	20	
		Quadruple	340	70	20	
SLS (Switzerland)	Booster 2.7 GeV 3 Hz	Dipole	950	1000	100	10
		Quadruple	140	120	100	100
Soleil (France)	Booster 2.75 GeV 3 Hz	Dipole	580	1000	50	50
		Quadruple	250	450	50	50
LNLS (Brazil)	Booster 500 MeV 10 Hz	Dipole	300	420	10	120
		Quadruple	10	21	10	10

described in earlier chapter that the power saving in White circuit is huge in comparison to non-White circuit for which, a strong coupling of primary winding with the energy storage choke is essential to improve the performance. This made the selection of White circuit for magnet excitation as the preferred choice for decades in which an independent d.c. supply is added to act as a bias to the a.c. sine waveform designed to feed the unidirectional magnetic field to reduce the loads on the mains supply. The switched mode converter with IGBT made a significant contribution in introducing a long-awaited alternative technology with the choice of new topology to control very high conducting currents in both the directions and the reasons for its preference has been discussed. Converter topologies for supplying power to the accelerator magnets are sometimes modified or upgraded to take full advantage of the advanced technology during the passage of time. For example, the comparison of RAL (ISIS, UK), 800 MeV Spallation Neutron Source (SNS) installed in 1984, and 1.6 GeV Chinese Spallation Neutron Source (CSNS) reveals that, while RAL used rotating machine, 1 MVA, three phase motor–alternator for generating sine wave (50 Hz) to supply makeup power connected to energy storage choke of White circuit through a complex scheme of conversion from three phase to single phase at the initial stage which was subsequently upgraded as the technology advanced, CSNS (Chinese Spallation Neutron source) adopted semiconductor-based converter topology for the generation of 25 Hz a.c. supply with PWM to the energy storage choke in 2004. Besides, it has fully utilized the available advanced technology in designing power converter modules with IGBTs connected in parallel to supply the desired current for booster (Qi et al. 2004). There are a number of such examples to establish that variations are primarily on the timing of the project, availability of technology and its adaptation, requirements of magnet physics, orientation of magnet cells, improvement in performance, operation, maintenance facilities, the cycling frequency, and size of accelerator or combination of accelerators. It is a hard task to review and analyze a large number of converter topologies from the vast world of operational accelerators specifically, with the converters applicable in capacitive type energy storage or for direct supply connection to accelerators. In spite of much difficulty, a large number of accelerators are reviewed in Visintini (2004) which contributed in depth analysis and outlined various considerations in designing power supplies for particle accelerators. As mentioned several times that comparison of initial synchrotron accelerators was possible because their system configuration was more or less common. But, it is difficult to draw similar comparison for present-day accelerators because the type of magnet excitation, the choice of semiconductor devices, converter topologies, and regulator selections are different for different machines. However, in spite of such differences, the adopted technology compiled from the available technical literatures and presented in Table 7.4 tried to strengthen the similarities and establish that the variation in output current, which is the most important parameter of the converter topology, is acceptable only in "ppm" ranges for some widely used operational accelerators.

The importance of cycling frequency and ramping type at the initial stage of the design can be best explained by considering the bending magnet power supply of the most prominent particle accelerator, SLS (Swiss Light Source) as an example.

Table 7.4 Adopted power supply technology in different projects

Sr. no.	Project	Linac (MeV)	Frequency (Hz)	Booster (GeV)	Storage ring (GeV)	Adopted technology
1.	Diamond (UK)	100	5	3.0	3.0	1000 d.c. power converters; PWM technology with digital regulation
2.	LNLS (Brazil)	120	3	0.50	1.37	200 power converter (d.c. and a.c) adopted thyristor bridge and PWM technology
3.	ESRF (France)		10	3		White circuit, d.c. fixed link, H bridge with GTO
4.	SLS (Switzerland)	100	3	2.75		640 d.c. converters, fully digital regulation and control
5.	CSNS (China)	80	25	1.6		SCR rectifier, d.c-d.c chopper and H-type PWM inverter
6.	Soleil (France)	100	3	2.75		12 pulse bridge with PWM
7.	SSRL (USA)	120	10	3		White circuit—12 pulse SCR 2 × 480 kW—voltage regulated
8a.	DESY-II DORIS (Germany)		12.5	4.5		White circuit is divided into 12 sections, 12 pulse SCR supply
8b.	DESY-II PETRA (Germany)		12.5	7		PWM, mixed analog and digital regulations
9.	BESSY-II Berlin (Germany)		10	1.9		Switched mode PS IGBT working in H bridge

This has been referred earlier several times which influenced in the selection of power supplies for Diamond (UK) and other accelerators. In this case, magnet represents a load of 0.6 Ω and 80 mH and the design specification is for a peak current of 950 A at a maximum ramping frequency of 3 Hz. This results a current slew of 10 kA/s and 1000 V peak voltage requirement. An average of 205 kW is required to power the dipole magnets for sinusoidal operation. The power swing between +750 and –250 kW explains the stringent requirements (Irminger et al. 1998). Diamond and other accelerators also exhibit similar characteristics. The pattern of magnet excitation may change but typical magnet current/voltage characteristics and its resultant, the power swing with the type of ramping remained the same as shown earlier in case of NIMROD of Rutherford Laboratory for triangular ramping which can be comparable. To follow the established practice and ease the selection, such analysis and comparisons with other operating synchrotrons are always important for future projects.

Review of Control Systems

So far, the different practice of magnet excitations has been discussed and compared their strength and inadequacies, but it is even more difficult to compare on the structure of measurements and control system as it is different for different accelerator. Besides, the control strategies are defined and fixed as per the availability of control components during the time of the project stage. Advancement in measurements and control has similar history like the changing pattern of rectifiers from Hg–arc to semiconductor devices like IGBT or beyond. It started with the pneumatic controller and currently digital processing has taken over. In most manufacturing and process industries, even now, pneumatic controllers are popular and these controllers operate with standardized inputs and outputs with air pressure ranging between 3 and 15 psi and generate control signals which are combination of proportional, integral, and derivative actions, in short, termed as PID controllers. The drawback of the pneumatic system lied with the stability of the air pressure and its dryness factor. Over the years, the transfer from traditional pneumatic to electronic system was slowly initiated. This transition phase was a trying time to introduce and adopt the facilities offered by the electronic system wherever possible to improve accuracy of measurement and control of process parameters. So, the pneumatic signal values of 3–15 psi with its equivalent electronic signal values of 4–20 mA continued for indication, recording and control of field instruments to maintain the process requirements. PID controllers are extensively used in process industries. In fact, in most cases the desired controller structure is selected from a menu of the available controllers which are any of the following three types of standard PID configurations. Proportional (P), Proportional + Integral (PI), and Proportional + Integral + Derivative (PID) are available for selection to achieve the desired control responses (Dey 2020). The algorithm of a PID control loop configuration is shown below:

$$C(t) = \frac{1}{K_p}\left[e(t) + \frac{1}{T_i}\int ed(t) + T_d\left(\frac{de}{dt}\right) \right]$$

Steady-state gain = $\dfrac{1}{K_p}$

System error = $e(t)$
Integral action time = T_i
Derivative action time = T_d

In many applications, it is possible to achieve acceptable performance by the use of traditional and rule of thumb tuning methods. Such procedures were quite time consuming and did not always give the best possible result because control functions especially for process control with large settling time offers some unexpected and complicated behavior. Furthermore, the controller settings designed for one particular operating state may be inappropriate for other operating conditions. In some situations, the physical principles may not be well understood or the necessary process parameters cannot be measured. When this happens, it will not be possible to obtain relativistic mathematical models. The journey from traditional pneumatic instruments in control applications to total digital system in last 60 years is a glorious achievement, and this is particularly laudable because improvement in performance of control function has increased many times. As per the requirement, due emphasis is placed on improving and selecting the quality of sensing devices, methods, and integrity in measurements with respect to achieving higher accuracy. In justifying the choice of a suitable control systems for accepting automation for a particular application including close monitoring on the functioning of the final control elements by installing an effective instrument loop to obtain the benefits that accrued on such costly investments which pays large dividend to the users. It is necessary for all the process industries to develop a suitable structure for instrumentation to control complex process variables which may be linear, nonlinear, and time-variant or time-invariant in nature. In developing a structure, the instrumentation setup in any industry comprises few important process parameters which are usually common to most production units. These common features are mainly temperature, pressure, flow, level, pH, humidity, position, speed, weight, and other parameters. The only difference in the selection of instruments depends on the type of industry, the acceptable range of measurement, the accuracy demanded, the visibility of historical data, the method of control mechanism to be adopted, either predictive or adaptive, the involvement of human-machine interface (HMI), and the choice of final control elements are just a few mentioned here. But now, since the buzz word for automation is the accuracy in all the areas, such as in sensing, measurements, metering, transporting of signal from remote site, control and actuation of final control elements with minimum loss of signal in transmission, efforts are always made to minimize errors in all aspects to get the best result. In some situations, the physical principles may not be well understood or the necessary process parameters cannot be measured and when such situation arises, it will not be possible to set up realistic mathematical models. For dynamic industrial process, it is better to split the control systems to layers so that a follow-up of each control layer is possible to track the signal of the chosen process variables. In this way, design of optimum control layers is possible and desired set point values are maintained. For most applications, multivariable process can adequately be analyzed by dividing the system into a number of sub-systems. In accelerator operation unlike process or manufacturing industries, the parameters are very few, but their importance are vital for operating the accelerator which invites sensing and measurements of flux density, vacuum, cooling water temperatures, radiation monitoring, current control in r.f. generators, in different magnet windings to the highest order to keep up the level of desired performance of the facilities. To achieve improved performance and reasonable beam intensities as per the design expectation at comparable costs, the application of knowledge-based system technique in the domain of computer control is emerging as a powerful tool to the engineers working in accelerator projects. With existing dedicated instrumentation facilities specific stability demand of current inadequacies with effective monitoring and control has been highly improved. Expert planning system is basically knowledge and rule based (Dey 1990). The knowledge gained from experiences of accelerator operation for years, set rules are almost framed and expert operators are familiar with the system to prompt and suggest in developing a program to follow the sequences desired for the particle accelerator operation. The identification and the interactive nature of the following key values in structure development are dominant to represent an expert system model:

- Identification and measurement of current requirements of dipole, quadrupole, sextupole, and other magnets.
- Follow the exact program sequence and the time frame at each stage of accelerator operation.
- Develop quality current sensors for selection in control loop structures and validate accuracy tests as required by controller algorithms.
- Control is essentially an important factor in implementing expert planning system successfully to user's satisfaction of facilities offered.
- The irregularities on control can develop when the system represented in the usual form is not proper due to
  - Deficiency in the form or content of accelerator operational activities
  - Unpredictability of the changing nature or non-linearity of magnet circuits

Continuous plants with multivariable process parameters create large-scale control problem and requires long settling time which may vitiates the result. Usually, how many measurements are to be made and where the measurement points are to be located for reliability of measurements and transmission of signals is an important factor in the selection with regard to cost and reliability and linking data from remote site to a central control station. In practice, accurate models of industrial processes are rarely available and therefore, the algorithms are to be developed based on process knowledge and field experience over the years in handling the particular process. Careful design in the identification of tasks and for optimization a rule-based supervisory algorithm is needed to present the operating constraints imposed by the design of plant.

Summing Up

It is essential to follow a minimum functional structure for identifying the accountability of different disciplines and monitoring of activities minutely for a large accelerator project. The importance of semiconductor devices, especially high power thyristors, power diodes, converter topologies, and application of capacitors as storage devices needs to be thoroughly explored before the execution of the project. The regulation in older synchrotron machines was analog, but digital regulation is becoming popular and increasing adoption of digital control systems with sophisticated transducers are considered to be the basis of future control system strategy. Present-day accelerators are computer controlled and the performance depends on the accuracy and the reliability of output sensing transducers, the ADC and DAC and their effective utilization in the feedback loop.

The scientific research and technological innovations are to move in parallel to transport our dreams and aspirations to new heights and serve for the betterment of the society. It all started hundred years back from now with Lord Rutherford's epoch making discoveries since then; noteworthy achievements are made in all disciplines. Let the dedication of so many scientists, engineers, and technologists be honored by continuing to cross new boundaries in future assignments and expect to achieve the highest peak at all fronts with the spirit of the pioneers.

References

Dey PK (1990) Expert planning system for copper smelter. In: Conference on expert planning systems. IEE, London, p 322
Dey PK (2020) Measurements and controls—some case studies with copper production. J Inst Eng (India) 101:317
Dobbling JA (2006) Diamond booster magnet power converters
Irminger G et al (1998) A 3 Hz, 1 MW peak bending magnet power supply for the Swiss Light Source. PSI, Villigen
Marks N (1996) The choice of power converter systems for a 3 GeV booster synchrotron
Qi X et al (2004) Resonant magnet power supply system for the rapid cycle synchrotron of Chinese spallation neutron source. In: APAC 2004, Gyeongju, Korea
Visintini R (2004) Power converters for accelerators. CERN Accelerator School (CAS), Warrington

GPSR Compliance

The European Union's (EU) General Product Safety Regulation (GPSR) is a set of rules that requires consumer products to be safe and our obligations to ensure this.

If you have any concerns about our products, you can contact us on ProductSafety@springernature.com

In case Publisher is established outside the EU, the EU authorized representative is:

Springer Nature Customer Service Center GmbH
Europaplatz 3
69115 Heidelberg, Germany

Batch number: 10704307

Printed by Printforce, the Netherlands